STUDYING MADE EASY

This Cram101 notebook is designed to make studying easier and increase your comprehension of the textbook material. Instead of starting with a blank notebook and trying to write down everything discussed in class lectures, you can use this Cram101 textbook notebook and annotate your notes along with the lecture.

Our goal is to give you the best tools for success.

For a supreme understanding of the course, pair your notebook with our online tools. Should you decide you prefer Cram101.com as your study tool,

we'd like to offer you a trade...

Our Trade In program is a simple way for us to keep our promise and provide you the best studying tools, regardless of where you purchased your Cram101 textbook notebook. As long as your notebook is in *Like New Condition**, you can send it back to us and we will immediately give you a Cram101.com account free for 120 days!

Let The Trade In Begin!

THREE SIMPLE STEPS TO TRADE:

1. Go to www.cram101.com/tradein and fill out the packing slip information.

2. Submit and print the packing slip and mail it in with your Cram101 textbook notebook.

3. Activate your account after you receive your email confirmation.

* Books must be returned in *Like New Condition*, meaning there is no damage to the book including, but not limited to; ripped or torn pages, markings or writing on pages, or folded / creased pages. Upon receiving the book, Cram101 will inspect it and reserves the right to terminate your free Cram101.com account and return your textbook notebook at the owners expense.

Visit Cram101.com for full Practice Exams

"Just the Facts101" is a Cram101 publication and tool designed to give you all the facts from your textbooks. Visit Cram101.com for the full practice test for each of your chapters for virtually any of your textbooks.

Cram101 has built custom study tools specific to your textbook. We provide all of the factual testable information and unlike traditional study guides, we will never send you back to your textbook for more information.

YOU WILL NEVER HAVE TO HIGHLIGHT A BOOK AGAIN!

Cram101 StudyGuides
All of the information in this StudyGuide is written specifically for your textbook. We include the key terms, places, people, and concepts... the information you can expect on your next exam!

Want to take a practice test?
Throughout each chapter of this StudyGuide you will find links to cram101.com where you can select specific chapters to take a complete test on, or you can subscribe and get practice tests for up to 12 of your textbooks, along with other exclusive cram101.com tools like problem solving labs and reference libraries.

Cram101.com
Only cram101.com gives you the outlines, highlights, and PRACTICE TESTS specific to your textbook. Cram101.com is an online application where you'll discover study tools designed to make the most of your limited study time.

By purchasing this book, you get 50% off the normal monthly subscription fee!. Just enter the promotional code **'DK73DW20492'** on the Cram101.com registration screen.

www.Cram101.com

Learning System

Just The

facts101

Textbook Key Facts

Textbook Outlines, Highlights, and Practice Quizzes

Linear Programming and Network Flows

by Mokhtar S. Bazaraa, John J. Jarvis, Hanif D. Sherali, 4th Edition

All "Just the Facts101" Material Written or Prepared by Cram101 Publishing

Title Page

Linear Programming and Network Flows
Mokhtar S. Bazaraa, John J. Jarvis, Hanif D. Sherali, 4th

CONTENTS

CHAPTER OUTLINE: KEY TERMS, PEOPLE, PLACES, CONCEPTS

Breakthrough

Flow

Matrix

Discrete optimization

Slack variable

Linear approximation

Optimization

Mathematical model

Assignment

Assignment problem

Optimal control

Cutting stock problem

Capital budgeting

Space

Bounded set

Notation

Facility location

Selection

Artificial intelligence

Chapter 1. INTRODUCTION

Breakthrough	Breakthrough is an abstract strategy board game invented by Dan Troyka in 2000 and made available as a Zillions of Games file (ZRF). It won the 2001 8x8 Game Design Competition, even though the game was originally played on a 7x7 board, as it is trivially extendible to larger board sizes. Rules The board is initially set up as shown on the right.
Flow	Flow is middleware software, which allows data integration specialists to connect disparate systems, whether they are on-premise, hosted or in the cloud; transforming and restructuring data as required between environments. Flow functionality can be utilised for data integration projects, EDI and data conversion activities. Flow has been created by Flow Software Ltd in NZ and is available through a variety of partner companies or directly from Flow Software in NZ and Australia.
Matrix	In hot metal typesetting, a matrix is a mold for casting the letters known as sorts used in letterpress printing. In letterpress typography the matrix of one letter is inserted into the bottom of a hand mould, the mould is locked and molten type metal is poured into a straight-sided vertical cavity above the matrix. When the metal has cooled and solidified the mould is unlocked and a newly-cast metal sort is removed, ready for composition with other sorts.
Discrete optimization	Discrete optimization is a branch of optimization in applied mathematics and computer science. As opposed to continuous optimization, the variables used in the mathematical program are restricted to assume only discrete values, such as the integers. Two notable branches of discrete optimization are:•combinatorial optimization, which refers to problems on graphs, matroids and other discrete structures•integer programming These branches are closely intertwined however since many combinatorial optimization problems can be modeled as integer programs (e.g. shortest path) and conversely, integer programs can often be given a combinatorial interpretation.
Slack variable	In an optimization problem, a slack variable is a variable that is added to an inequality constraint to transform it to an equality. Introducing a slack variable replaces an inequality constraint with an equality constraint and a nonnegativity constraint.

Linear approximation	In mathematics, a linear approximation is an approximation of a general function using a linear function (more precisely, an affine function). They are widely used in the method of finite differences to produce first order methods for solving or approximating solutions to equations.

Given a twice continuously differentiable function f of one real variable, Taylor's theorem for the case n = 1 states that $f(x) = f(a) + f'(a)(x - a) + R_2$

where R_2 is the remainder term. |
| Optimization | In mathematics, computer science and economics, optimization, refers to choosing the best element from some set of available alternatives.

In the simplest case, this means solving problems in which one seeks to minimize or maximize a real function by systematically choosing the values of real or integer variables from within an allowed set. This formulation, using a scalar, real-valued objective function, is probably the simplest example; the generalization of optimization theory and techniques to other formulations comprises a large area of applied mathematics. |
| Mathematical model | A mathematical model is a description of a system using mathematical concepts and language. The process of developing a mathematical model is termed mathematical modelling. Mathematical models are used not only in the natural sciences (such as physics, biology, earth science, meteorology) and engineering disciplines (e.g. computer science, artificial intelligence), but also in the social sciences (such as economics, psychology, sociology and political science); physicists, engineers, statisticians, operations research analysts and economists use mathematical models most extensively. |
| Assignment | In computer programming, an assignment statement sets or re-sets the value stored in the storage location(s) denoted by a variable name. In most imperative computer programming languages, assignment statements are one of the basic statements. Common notations for the assignment operator are and . |
| Assignment problem | The assignment problem is one of the fundamental combinatorial optimization problems in the branch of optimization or operations research in mathematics. It consists of finding a maximum weight matching in a weighted bipartite graph.

In its most general form, the problem is as follows:There are a number of agents and a number of tasks. |
| Optimal control | Optimal control theory, an extension of the calculus of variations, is a mathematical optimization method for deriving control policies. The method is largely due to the work of Lev Pontryagin and his collaborators in the Soviet Union and Richard Bellman in the United States. |

Chapter 1. INTRODUCTION

Cutting stock problem	The cutting-stock problem is an optimization problem, or more specifically, an integer linear programming problem. It arises from many applications in industry. Imagine that you work in a paper mill and you have a number of rolls of paper of fixed width waiting to be cut, yet different customers want different numbers of rolls of various-sized widths. How are you going to cut the rolls so that you minimize the waste (amount of left-overs)?

Solving this problem to optimality can be economically significant: a difference of 1% for a modern paper machine can be worth more than one million USD per year. Formulation and solution approaches

The standard formulation for the cutting-stock problem (but not the only one) starts with a list of m orders, each requiring q_j, j = 1,..m pieces. We then construct a list of all possible combinations of cuts (often called 'patterns'), associating with each pattern a positive integer variable x_i representing how many times each pattern is to be used. The linear integer program is

then:minimize $\sum_{i=1}^{n} c_i x_i$ subject to $\sum_{i=1}^{n} a_{ij} x_i \geq q_j, \qquad \forall j = 1, \ldots, m$ and

$x_i \geq 0$, integer

where a_{ij} is the number of times order j appears in pattern and c_i is the cost (often the waste) of pattern . The precise nature of the quantity constraints can lead to subtly different mathematical characteristics. The above formulation's quantity constraints are minimum constraints (at least the given amount of each order must be produced, but possibly more). When c_i = 1 the objective minimises the number of utilised master items and, if the constraint for the quantity to be produced is replaced by equality, it is called the bin packing problem. The most general formulation has two-sided constraints (and in this case a minimum-waste solution may consume more than the

$$q_j \leq \sum_{i=1}^{n} a_{ij} x_i \leq Q_j, \qquad \forall j = 1, \ldots, m$$

minimum number of master items):

This formulation applies not just to one-dimensional problems. Many variations are possible, including one where the objective is not to minimise the waste, but to maximise the total value of the produced items, allowing each order to have a different value.

In general, the number of possible patterns grows exponentially as a function of m, the number of orders. As the number of orders increases, it may therefore become impractical to enumerate the possible cutting patterns.

An alternative approach uses delayed column-generation. This method solves the cutting-stock problem by starting with just a few patterns. It generates additional patterns when they are needed.

For the one-dimensional case, the new patterns are introduced by solving an auxiliary optimization problem called the knapsack problem, using dual variable information from the linear program. The knapsack problem has well-known methods to solve it, such as branch and bound and dynamic programming. The Delayed Column Generation method can be much more efficient than the original approach, particularly as the size of the problem grows. The column generation approach as applied to the cutting stock problem was pioneered by Gilmore and Gomory in a series of papers published in the 1960s.

Capital budgeting

Capital budgeting is the planning process used to determine whether an organization's long term investments such as new machinery, replacement machinery, new plants, new products, and research development projects are worth pursuing. It is budget for major capital, or investment, expenditures.

Many formal methods are used in capital budgeting, including the techniques such as•Accounting rate of return•Net present value•Profitability index•Internal rate of return•Modified internal rate of return•Equivalent annuity

These methods use the incremental cash flows from each potential investment, or project.

Space

In mathematics, a space is a set with some added structure.

Mathematical spaces often form a hierarchy, i.e., one space may inherit all the characteristics of a parent space. For instance, all inner product spaces are also normed vector spaces, because the inner product induces a norm on the inner product space such that: $\|x\| = \sqrt{\langle x, x \rangle}.$

Modern mathematics treats 'space' quite differently compared to classical mathematics.

Bounded set

In mathematical analysis and related areas of mathematics, a set is called bounded, if it is, in a certain sense, of finite size. Conversely, a set which is not bounded is called unbounded. The word bounded makes no sense in a general topological space, without a metric.

A set S of real numbers is called bounded from above if there is a real number k such that k ≥ s for all s in S. The number k is called an upper bound of S. The terms bounded from below and lower bound are similarly defined.

A set S is bounded if it has both upper and lower bounds. Therefore, a set of real numbers is bounded if it is contained in a finite interval. Metric space

A subset S of a metric space (M, d) is bounded if it is contained in a ball of finite radius, i.e. if there exists x in M and r > 0 such that for all s in S, we have d(x, s) < r. M is a bounded metric space (or d is a bounded metric) if M is bounded as a subset of itself.

•Total boundedness implies boundedness. For subsets of R^n the two are equivalent.•A metric space is compact if and only if it is complete and totally bounded.•A subset of Euclidean space R^n is compact if and only if it is closed and bounded.Boundedness in topological vector spaces

In topological vector spaces, a different definition for bounded sets exists which is sometimes called von Neumann boundedness.

Notation	The term notation can refer to: •Phonographic writing systems, by definition, use symbols to represent components of auditory language, i.e. speech, which in turn refers to things or ideas. The two main kinds of phonographic notational system are the alphabet and syllabary. Some written languages are more consistent in their correlation of written symbol or grapheme and sound or phoneme, and are therefore considered to have better phonemic orthography.•Ideographic writing, by definition, refers to things or ideas independently of their pronunciation in any language.
Facility location	Facility location, is a branch of operations research and computational geometry concerning itself with mathematical modeling and solution of problems concerning optimal placement of facilities in order to minimize transportation costs, avoid placing hazardous materials near housing, outperform competitors' facilities, etc.

A simple facility location problem is the Fermat-Weber problem, in which a single facility is to be placed, with the only optimization criterion being the minimization of the sum of distances from a given set of point sites. More complex problems considered in this discipline include the placement of multiple facilities, constraints on the locations of facilities, and more complex optimization criteria.

Selection	In relational algebra, a selection (sometimes called a restriction to avoid confusion with SQL's use of SELECT) is a unary operation written as $\sigma_{a\theta b}(R)$ or $\sigma_{a\theta v}(R)$ where:• a and b are attribute names• θ is a binary operation in the set $\{<, \leq, =, \geq, >\}$.• v is a value constant• R is a relation

The selection $\sigma_{a\theta b}(R)$ selects all those tuples in R for which θ holds between the a and the b attribute.

The selection $\sigma_{a\theta v}(R)$ selects all those tuples in R for which θ holds between the a attribute and the value v.

For an example, consider the following tables where the first table gives the relation $Person$, the second table gives the result of $\sigma_{Age \geq 34}(Person)$ and the third table gives the result of

Chapter 1. INTRODUCTION

CHAPTER HIGHLIGHTS & NOTES: KEY TERMS, PEOPLE, PLACES, CONCEPTS

Artificial intelligence	Artificial intelligence is the intelligence of machines and the branch of computer science that aims to create it. artificial intelligence textbooks define the field as 'the study and design of intelligent agents' where an intelligent agent is a system that perceives its environment and takes actions that maximize its chances of success. John McCarthy, who coined the term in 1956, defines it as 'the science and engineering of making intelligent machines.'

CHAPTER QUIZ: KEY TERMS, PEOPLE, PLACES, CONCEPTS

1. _____, is a branch of operations research and computational geometry concerning itself with mathematical modeling and solution of problems concerning optimal placement of facilities in order to minimize transportation costs, avoid placing hazardous materials near housing, outperform competitors' facilities, etc. Minsum _____

   A simple _____ problem is the Fermat-Weber problem, in which a single facility is to be placed, with the only optimization criterion being the minimization of the sum of distances from a given set of point sites. More complex problems considered in this discipline include the placement of multiple facilities, constraints on the locations of facilities, and more complex optimization criteria.

   a. Fixes that fail
   b. Flow network
   c. Facility location
   d. FuncDesigner

2. The _____ is one of the fundamental combinatorial optimization problems in the branch of optimization or operations research in mathematics. It consists of finding a maximum weight matching in a weighted bipartite graph.

   In its most general form, the problem is as follows: There are a number of agents and a number of tasks.

   a. Assignment problem
   b. Integer points in convex polyhedra
   c. Modulo
   d. Necessity and sufficiency

3. . A _____ is a description of a system using mathematical concepts and language. The process of developing a _____ is termed mathematical modelling. _____s are used not only in the natural sciences (such as physics, biology, earth science, meteorology) and engineering disciplines (e.g. computer science, artificial intelligence), but also in the social sciences (such as economics, psychology, sociology and political science); physicists, engineers, statisticians, operations research analysts and economists use _____s most extensively.

a. Matrix decomposition

b. Metatheorem

c. Mathematical model

d. Necessity and sufficiency

4. In relational algebra, a _____(sometimes called a restriction to avoid confusion with SQL's use of SELECT) is a unary operation written as $\sigma_{a\theta b}(R)$ or $\sigma_{a\theta v}(R)$ where:• a and b are attribute names• θ is a binary operation in the set $\{<, \leq, =, \geq, >\}$. v is a value constant• R is a relation

The _____ $\sigma_{a\theta b}(R)$ selects all those tuples in R for which θ holds between the a and the b attribute.

The _____ $\sigma_{a\theta v}(R)$ selects all those tuples in R for which θ holds between the a attribute and the value v.

For an example, consider the following tables where the first table gives the relation $Person$, the second table gives the result of $\sigma_{Age \geq 34}(Person)$ and the third table gives the result of $\sigma_{Age=Weight}(Person)$.

a. .NET Messenger Service

b. Selection

c. Frequency partition of a graph

d. Friendship paradox

5. _____ is an abstract strategy board game invented by Dan Troyka in 2000 and made available as a Zillions of Games file (ZRF). It won the 2001 8x8 Game Design Competition, even though the game was originally played on a 7x7 board, as it is trivially extendible to larger board sizes.

Rules

The board is initially set up as shown on the right.

a. Breakthru

b. Camelot

c. Capture Go

d. Breakthrough

ANSWER KEY
Chapter 1. INTRODUCTION

1. c
2. a
3. c
4. b
5. d

You can take the complete Chapter Practice Test

for Chapter 1. INTRODUCTION
on all key terms, persons, places, and concepts.

Online 99 Cents

http://www.epub7.6.20492.1.cram101.com/

Use www.Cram101.com for all your study needs

including Cram101's online interactive problem solving labs in

chemistry, statistics, mathematics, and more.

CHAPTER OUTLINE: KEY TERMS, PEOPLE, PLACES, CONCEPTS

	Convex analysis
	Row vector
	Coordinate vector
	Matrix
	Space
	Linear subspace
	Independence
	SubSpace
	Symmetric matrix
	Transposition
	ELEMENTARY
	Diagonal matrix
	Adjoint
	Dual problem
	Convex set
	Convex combination
	Recession cone
	Direction vector
	Concave function

	Redundancy
	Assignment
	Assignment problem
	Order
	Bounded set
	Representation theorem
	Open set
	Supporting hyperplane

CHAPTER HIGHLIGHTS & NOTES: KEY TERMS, PEOPLE, PLACES, CONCEPTS

Convex analysis	Convex analysis is the branch of mathematics devoted to the study of properties of convex functions and convex sets, often with applications in convex minimization, a subdomain of optimization theory.

A convex set is a set $C \subseteq X$, for some vector space X, such that for any $x, y \in C$ and $\lambda \in [0, 1]$ then $\lambda x + (1 - \lambda)y \in C$. Convex functions

A convex function is any extended real-valued function $f : X \to \mathbb{R} \cup \{\pm\infty\}$ which satisfies Jensen's inequality, i.e. for any $x, y \in X$ and any $\lambda \in [0, 1]$ then
$$f(\lambda x + (1 - \lambda)y) \leq \lambda f(x) + (1 - \lambda)f(y).$$

Equivalently, a convex function is any (extended) real valued function such that its epigraph
$$\{(x, r) \in X \times \mathbb{R} : f(x) \leq r\}$$

is a convex set. Convex conjugate

The convex conjugate of an extended real-valued (not necessarily convex) function $f : X \to \mathbb{R} \cup \{\pm\infty\}$ is $f^* : X^* \to \mathbb{R} \cup \{\pm\infty\}$ where X^* is the dual space of X, and

$$f^*(x^*) = \sup_{x \in X}\{\langle x^*, x \rangle - f(x)\}$$

:pp.75-79 Biconjugate

The biconjugate of a function $f : X \to \mathbb{R} \cup \{\pm\infty\}$ is the conjugate of the conjugate, typically written as $f^{**} : X \to \mathbb{R} \cup \{\pm\infty\}$.

Row vector

In linear algebra, a row vector is a 1 × n matrix, that is, a matrix consisting of a single row:

$$\mathbf{x} = \begin{bmatrix} x_1 & x_2 & \ldots & x_m \end{bmatrix}.$$

$$\begin{bmatrix} x_1 & x_2 & \ldots & x_m \end{bmatrix}^{\mathrm{T}} = \begin{bmatrix} x_1 \\ x_2 \\ \vdots \\ x_m \end{bmatrix}.$$

The transpose of a row vector is a column vector:

The set of all row vectors forms a vector space which acts like the dual space to the set of all column vectors, in the sense that any linear functional on the space of column vectors (i.e. any element of the dual space) can be represented uniquely as a dot product with a specific row vector.

Notation

Row vectors are sometimes written using the following non-standard notation:

$$\mathbf{x} = \begin{bmatrix} x_1, x_2, \ldots, x_m \end{bmatrix}.$$

Operations•Matrix multiplication involves the action of multiplying each row vector of one matrix by each column vector of another matrix.•The dot product of two vectors a and b is equivalent to multiplying the row vector representation of a by the column vector representation of b:

$$\mathbf{a} \cdot \mathbf{b} = \begin{bmatrix} a_1 & a_2 & a_3 \end{bmatrix} \begin{bmatrix} b_1 \\ b_2 \\ b_3 \end{bmatrix}.$$

Preferred input vectors for matrix transformations

Frequently a row vector presents itself for an operation within n-space expressed by an n by n matrix M:v M = p.

Then p is also a row vector and may present to another n by n matrix Q:p Q = t.

Conveniently, one can write t = p Q = v MQ telling us that the matrix product transformation MQ can take v directly to t. Continuing with row vectors, matrix transformations further reconfiguring n-space can be applied to the right of previous outputs.

Coordinate vector	In linear algebra, a coordinate vector is an explicit representation of a vector in an abstract vector space as an ordered list of numbers or, equivalently, as an element of the coordinate space F^n. Coordinate vectors allow calculations with abstract objects to be transformed into calculations with blocks of numbers (matrices and column vectors).
Matrix	In hot metal typesetting, a matrix is a mold for casting the letters known as sorts used in letterpress printing.
	In letterpress typography the matrix of one letter is inserted into the bottom of a hand mould, the mould is locked and molten type metal is poured into a straight-sided vertical cavity above the matrix. When the metal has cooled and solidified the mould is unlocked and a newly-cast metal sort is removed, ready for composition with other sorts.
Space	In mathematics, a space is a set with some added structure.
	Mathematical spaces often form a hierarchy, i.e., one space may inherit all the characteristics of a parent space. For instance, all inner product spaces are also normed vector spaces, because the inner product induces a norm on the inner product space such that: $\|x\| = \sqrt{\langle x, x \rangle}$.
	Modern mathematics treats 'space' quite differently compared to classical mathematics.
Linear subspace	The concept of a linear subspace is important in linear algebra and related fields of mathematics. A linear subspace is usually called simply a subspace when the context serves to distinguish it from other kinds of subspaces.
	Let K be a field (such as the field of real numbers), and let V be a vector space over K. As usual, we call elements of V vectors and call elements of K scalars.
Independence	In probability theory, to say that two events are independent intuitively means that the occurrence of one event makes it neither more nor less probable that the other occurs.

For example:•The event of getting a 6 the first time a die is rolled and the event of getting a 6 the second time are independent.•By contrast, the event of getting a 6 the first time a die is rolled and the event that the sum of the numbers seen on the first and second trials is 8 are not independent.•If two cards are drawn with replacement from a deck of cards, the event of drawing a red card on the first trial and that of drawing a red card on the second trial are independent.•By contrast, if two cards are drawn without replacement from a deck of cards, the event of drawing a red card on the first trial and that of drawing a red card on the second trial are again not independent.

Similarly, two random variables are independent if the conditional probability distribution of either given the observed value of the other is the same as if the other's value had not been observed. The concept of independence extends to dealing with collections of more than two events or random variables.

SubSpace	SubSpace is a two-dimensional space shooter computer game published in 1997 by Virgin Interactive Entertainment (VIE) which was a finalist for the Academy of Interactive Arts & Sciences Online Game of the Year Award in 1998. This game, considered by some as freeware, and others as abandonware incorporates quasi-realistic zero-friction physics into a massively multiplayer online game. It is no longer operated by VIE; instead, fans and players of the game provide servers and technical updates. The action is viewed from above, which presents challenges very different from those of a three-dimensional game.
Symmetric matrix	In linear algebra, a symmetric matrix is a square matrix that is equal to its transpose. Let A be a symmetric matrix. Then: $A = A^\top$. The entries of a symmetric matrix are symmetric with respect to the main diagonal (top left to bottom right).
Transposition	In the methods of deductive reasoning in classical logic, transposition is the rule of inference that permits one to infer from the truth of 'A implies B' the truth of 'Not-B implies not-A', and conversely. Its symbolic expression is:$(P \rightarrow Q) \leftrightarrow (\sim Q \rightarrow \sim P)$ The '→' is the symbol for material implication and the doubleheaded arrow '↔' indicates a biconditional relationship. The symbol '~' indicates negation.
ELEMENTARY	In computational complexity theory, the complexity class ELEMENTARY of elementary recursive functions is the union of the classes in the exponential hierarchy. $$\begin{aligned} \text{ELEMENTARY} &= \text{EXP} \cup 2\text{EXP} \cup 3\text{EXP} \cup \cdots \\ &= \text{DTIME}(2^n) \cup \text{DTIME}(2^{2^n}) \cup \text{DTIME}(2^{2^{2^n}}) \cup \cdots \end{aligned}$$

The name was coined by László Kalmár, in the context of recursive functions and undecidability; most problems in it are far from elementary. Some natural recursive problems lie outside ELEMENTARY, and are thus NONELEMENTARY. Most notably, there are primitive recursive problems which are not in ELEMENTARY. We knowLOWER-ELEMENTARY $\subsetneqq$ EXPTIME $\subsetneqq$ ELEMENTARY $\subsetneqq$ PR

Whereas ELEMENTARY contains bounded applications of exponentiation (for example, $O\left(2^{2^n}\right)$), PR allows more general hyper operators (for example, tetration) which are not contained in ELEMENTARY.Definition

The definitions of elementary recursive functions are the same as for primitive recursive functions, except that primitive recursion is replaced by bounded summation and bounded product.

Diagonal matrix	In linear algebra, a diagonal matrix is a square matrix in which the entries outside the main diagonal are all zero. The diagonal entries themselves may or may not be zero.
Adjoint	In mathematics, the term adjoint applies in several situations. Several of these share a similar formalism: if A is adjoint to B, then there is typically some formula of the type(Ax,y) = (x, By).
	Specifically, adjoint may mean:•Hermitian adjoint in functional analysis•Adjoint functors in category theory•Adjoint representation of a Lie group•Adjoint endomorphism of a Lie algebra•Conjugate transpose of a matrix in linear algebra•Adjugate matrix, related to its inverse•Adjoint equation•The upper and lower adjoints of a Galois connection in order theory•For the adjoint of a differential operator with general polynomial coefficients see differential operator.
Dual problem	In constrained optimization, it is often possible to convert the primal problem (i.e. the original form of the optimization problem) to a dual form, which is termed a dual problem. Usually 'dual problem' refers to the 'Lagrangian dual problem' but other dual problems are used, for example, the Wolfe dual problem and the Fenchel dual problem. The Lagrangian dual problem is obtained by forming the Lagrangian, using nonnegative Lagrangian multipliers to add the constraints to the objective function, and then solving for some primal variable values that minimize the Lagrangian.
Convex set	In Euclidean space, an object is convex if for every pair of points within the object, every point on the straight line segment that joins them is also within the object. For example, a solid cube is convex, but anything that is hollow or has a dent in it, for example, a crescent shape, is not convex.
	The notion can be generalized to other spaces as described below. In vector spaces

Let S be a vector space over the real numbers, or, more generally, some ordered field. This includes Euclidean spaces. A set C in S is said to be convex if, for all x and y in C and all t in the interval [0,1], the point$(1 - t) x + t y$

is in C. In other words, every point on the line segment connecting x and y is in C. This implies that a convex set in a real or complex topological vector space is path-connected, thus connected.

Convex combination

In convex geometry, a convex combination is a linear combination of points (which can be vectors, scalars, or more generally points in an affine space) where all coefficients are non-negative and sum up to 1.

More formally, given a finite number of points $x_1, x_2, \ldots, x_n$ in a real vector space, a convex combination of these points is a point of the form $\alpha_1 x_1 + \alpha_2 x_2 + \cdots + \alpha_n x_n$

where the real numbers α_i satisfy $\alpha_i \geq 0$ and $\alpha_1 + \alpha_2 + \cdots + \alpha_n = 1.$

As a particular example, every convex combination of two points lies on the line segment between the points.

All convex combinations are within the convex hull of the given points.

Recession cone

In mathematics, especially convex analysis, the recession cone of a set A is a cone containing all vectors such that A recedes in that direction. That is, the set extends outward in all the directions given by the recession cone.

Given a nonempty set $A \subset X$ for some vector space X, then the recession cone $\mathrm{recc}(A)$ is given by $\mathrm{recc}(A) = \{y \in X : \forall x \in A, \forall \lambda \geq 0 : x + \lambda y \in A\}.$

If A is additionally a convex set then the recession cone can equivalently be defined by $\mathrm{recc}(A) = \{y \in X : \forall x \in A : x + y \in A\}.$

If A is a nonempty closed convex set then the recession cone can equivalently be defined as $\mathrm{recc}(A) = \bigcap_{t>0} t(A - a)$ for any choice of $a \in A.$ Properties •For any nonempty set A then $0 \in \mathrm{recc}(A)$.•For any nonempty convex set A then $\mathrm{recc}(A)$ is a convex cone.•For any nonempty closed convex set $A \subset X$ where

X is a finite dimensional Hausdorff space (e.g. $\mathbb{R}^d$), then $\mathrm{recc}(A) = \{0\}$ if and only if A is bounded. •For any nonempty convex set A then $A + \mathrm{recc}(A) = A$ where the sum is given by Minkowski addition. Relation to asymptotic cone

The asymptotic cone for $C \subseteq X$ is defined by
$$C_\infty = \{x \in X : \exists (t_i)_{i \in I} \subset (0, \infty), \exists (x_i)_{i \in I} \subset C : t_i \to 0, t_i x_i \to x\}.$$

By the definition it can easily be shown that $\mathrm{rec}(C) \subseteq C_\infty$.

In a finite dimensional space, then it can be shown that $C_\infty = \mathrm{rec}(C)$ if C is nonempty, closed and convex.

Direction vector	In mathematics, a direction vector that describes a line segment D is any vector $\overrightarrow{AB}$
	where A and B are two distinct points on the line D. If v is a direction vector for D, so is kv for any nonzero scalar k; and these are in fact all of the direction vectors for the line D. Under some definitions, the direction vector is required to be a unit vector, in which case each line has exactly two direction vectors, which are negatives of each other (equal in magnitude, opposite in direction).
	Direction vector for a line in R^2
	Any line in two-dimensional Euclidean space can be described as the set of solutions to an equation of the form ax + by + c = 0
	where a, b, c are real numbers. Then one direction vector of (D) is (− b,a).
Concave function	In mathematics, a concave function is the negative of a convex function. A concave function is also synonymously called concave downwards, concave down, convex upwards, convex cap or upper convex.
	A real-valued function f on an interval (or, more generally, a convex set in vector space) is said to be concave if, for any x and y in the interval and for any t in [0,1], $$f(tx + (1 - t)y) \geq tf(x) + (1 - t)f(y).$$
	A function is called strictly concave if $f(tx + (1 - t)y) > tf(x) + (1 - t)f(y)$ for any t in (0,1) and x ≠ y.

Redundancy	In engineering, redundancy is the duplication of critical components or functions of a system with the intention of increasing reliability of the system, usually in the case of a backup or fail-safe.
	In many safety-critical systems, such as fly-by-wire and hydraulic systems in aircraft, some parts of the control system may be triplicated, which is formally termed triple modular redundancy (TMR). An error in one component may then be out-voted by the other two.
Assignment	In computer programming, an assignment statement sets or re-sets the value stored in the storage location(s) denoted by a variable name. In most imperative computer programming languages, assignment statements are one of the basic statements. Common notations for the assignment operator are and .
Assignment problem	The assignment problem is one of the fundamental combinatorial optimization problems in the branch of optimization or operations research in mathematics. It consists of finding a maximum weight matching in a weighted bipartite graph.
	In its most general form, the problem is as follows:There are a number of agents and a number of tasks.
Order	Order (subtitled 'A Journal on the Theory of Ordered Sets and its Applications') is a quarterly peer-reviewed academic journal on order theory and its applications, published by Springer Science+Business Media. It was founded in 1984 by University of Calgary mathematics professor Ivan Rival; as of 2010, its editor in chief is Dwight Duffus, the Goodrich C. White Professor of Mathematics & Computer Science at Emory University and a former student of Rival's.
	According to the Journal Citation Reports, the 2009 impact factor of Order is 0.408, placing it in the fourth quartile of ranked mathematics journals.
Bounded set	In mathematical analysis and related areas of mathematics, a set is called bounded, if it is, in a certain sense, of finite size. Conversely, a set which is not bounded is called unbounded. The word bounded makes no sense in a general topological space, without a metric.
	A set S of real numbers is called bounded from above if there is a real number k such that k ≥ s for all s in S. The number k is called an upper bound of S. The terms bounded from below and lower bound are similarly defined.
	A set S is bounded if it has both upper and lower bounds. Therefore, a set of real numbers is bounded if it is contained in a finite interval. Metric space

A subset S of a metric space (M, d) is bounded if it is contained in a ball of finite radius, i.e. if there exists x in M and r > 0 such that for all s in S, we have d(x, s) < r. M is a bounded metric space (or d is a bounded metric) if M is bounded as a subset of itself. •Total boundedness implies boundedness. For subsets of R^n the two are equivalent.•A metric space is compact if and only if it is complete and totally bounded.•A subset of Euclidean space R^n is compact if and only if it is closed and bounded.Boundedness in topological vector spaces

In topological vector spaces, a different definition for bounded sets exists which is sometimes called von Neumann boundedness.

Representation theorem	In mathematics, a representation theorem is a theorem that states that every abstract structure with certain properties is isomorphic to a concrete structure.

For example,•in algebra, •Cayley's theorem states that every group is isomorphic to a transformation group on some set. Representation theory studies properties of abstract groups via their representations as linear transformations of vector spaces.•Stone's representation theorem for Boolean algebras states that every Boolean algebra is isomorphic to a field of sets.

Open set	In topology, a set U is called an open set if it is a neighborhood of every point $x \in U$. When dealing with metric spaces, this can be intuitively interpreted as saying that every $x \in U$ can be 'moved' some non-zero distance, in any direction, and it will still lie within U.

The notion of an open set provides a fundamental way to speak of nearness of points in a topological space, without explicitly having a concept of distance defined.

Supporting hyperplane	Supporting hyperplane is a concept in geometry. A hyperplane divides a space into two half-spaces. A hyperplane is said to support a set S in Euclidean space $\mathbb{R}^n$ if it meets both of the following:•S is entirely contained in one of the two closed half-spaces determined by the hyperplane•S has at least one point on the hyperplane.

Here, a closed half-space is the half-space that includes the hyperplane.

1. The concept of a _____ is important in linear algebra and related fields of mathematics. A _____ is usually called simply a subspace when the context serves to distinguish it from other kinds of subspaces.

   Let K be a field (such as the field of real numbers), and let V be a vector space over K. As usual, we call elements of V vectors and call elements of K scalars.

   a. Lyapunov vector
   b. Linear subspace
   c. Min-max theorem
   d. Mollifier

2. In mathematics, a _____ that describes a line segment D is any vector $\overrightarrow{AB}$

   where A and B are two distinct points on the line D. If v is a _____ for D, so is kv for any nonzero scalar k; and these are in fact all of the _____s for the line D. Under some definitions, the _____ is required to be a unit vector, in which case each line has exactly two _____s, which are negatives of each other (equal in magnitude, opposite in direction).

   _____ for a line in R^2

   Any line in two-dimensional Euclidean space can be described as the set of solutions to an equation of the formax + by + c = 0

   where a, b, c are real numbers. Then one _____ of (D) is (− b,a).

   a. Direction vector
   b. Four-vector
   c. Motion vector
   d. Poynting vector

3. In computer programming, an _____statement sets or re-sets the value stored in the storage location(s) denoted by a variable name. In most imperative computer programming languages, _____statements are one of the basic statements. Common notations for the assignment operator are and .

   a. Expression
   b. Identifier
   c. Assignment
   d. Reading

4. . _____ is the branch of mathematics devoted to the study of properties of convex functions and convex sets, often with applications in convex minimization, a subdomain of optimization theory.

A convex set is a set $C \subseteq X$, for some vector space X, such that for any $x, y \in C$ and $\lambda \in [0, 1]$ then $\lambda x + (1 - \lambda)y \in C$. Convex functions

A convex function is any extended real-valued function $f : X \to \mathbb{R} \cup \{\pm\infty\}$ which satisfies Jensen's inequality, i.e. for any $x, y \in X$ and any $\lambda \in [0, 1]$ then
$$f(\lambda x + (1 - \lambda)y) \leq \lambda f(x) + (1 - \lambda)f(y).$$

Equivalently, a convex function is any (extended) real valued function such that its epigraph
$$\{(x, r) \in X \times \mathbb{R} : f(x) \leq r\}$$

is a convex set. Convex conjugate

The convex conjugate of an extended real-valued (not necessarily convex) function $f : X \to \mathbb{R} \cup \{\pm\infty\}$ is $f^* : X^* \to \mathbb{R} \cup \{\pm\infty\}$ where X^* is the dual space of X, and
$$f^*(x^*) = \sup_{x \in X}\{\langle x^*, x \rangle - f(x)\}$$
:pp.75-79 Biconjugate

The biconjugate of a function $f : X \to \mathbb{R} \cup \{\pm\infty\}$ is the conjugate of the conjugate, typically written as $f^{**} : X \to \mathbb{R} \cup \{\pm\infty\}$.

a. Fritz John conditions
b. Convex analysis
c. Least absolute deviations
d. Mathematical programming with equilibrium constraints

5. In hot metal typesetting, a _____ is a mold for casting the letters known as sorts used in letterpress printing.

In letterpress typography the _____ of one letter is inserted into the bottom of a hand mould, the mould is locked and molten type metal is poured into a straight-sided vertical cavity above the _____. When the metal has cooled and solidified the mould is unlocked and a newly-cast metal sort is removed, ready for composition with other sorts.

a. Monitor proofing
b. Matrix
c. Pad printing
d. Paper density

ANSWER KEY
Chapter 2. LINEAR ALGEBRA, CONVEX ANALYSIS, AND POLYHEDRAL SETS

1. b
2. a
3. c
4. b
5. b

You can take the complete Chapter Practice Test

for Chapter 2. LINEAR ALGEBRA, CONVEX ANALYSIS, AND POLYHEDRAL SETS
on all key terms, persons, places, and concepts.

Online 99 Cents

http://www.epub7.6.20492.2.cram101.com/

Use www.Cram101.com for all your study needs

including Cram101's online interactive problem solving labs in

chemistry, statistics, mathematics, and more.

Lagrangian relaxation

Assignment

Assignment problem

Matrix

Degree

Degrees of freedom

Reduced cost

Freedom

Optimality criterion

Iteration

Block

PATH

Breakthrough

Exit

Reduced Vertical Separation Minima

Knapsack problem

Chapter 3. THE SIMPLEX METHOD

Lagrangian relaxation	In the field of mathematical optimization, Lagrangian relaxation is a relaxation method which approximates a difficult problem of constrained optimization by a simpler problem. A solution to the relaxed problem is an approximate solution to the original problem, and provides useful information.
	The method penalizes violations of inequality constraints using a Lagrangian multiplier, which imposes a cost on violations.
Assignment	In computer programming, an assignment statement sets or re-sets the value stored in the storage location(s) denoted by a variable name. In most imperative computer programming languages, assignment statements are one of the basic statements. Common notations for the assignment operator are and .
Assignment problem	The assignment problem is one of the fundamental combinatorial optimization problems in the branch of optimization or operations research in mathematics. It consists of finding a maximum weight matching in a weighted bipartite graph.
	In its most general form, the problem is as follows: There are a number of agents and a number of tasks.
Matrix	In hot metal typesetting, a matrix is a mold for casting the letters known as sorts used in letterpress printing.
	In letterpress typography the matrix of one letter is inserted into the bottom of a hand mould, the mould is locked and molten type metal is poured into a straight-sided vertical cavity above the matrix. When the metal has cooled and solidified the mould is unlocked and a newly-cast metal sort is removed, ready for composition with other sorts.
Degree	In mathematics, there are several meanings of degree depending on the subject.
	A degree (in full, a degree of arc, arc degree, or arcdegree), usually denoted by ° (the degree symbol), is a measurement of a plane angle, representing $\frac{1}{360}$ of a turn. When that angle is with respect to a reference meridian, it indicates a location along a great circle of a sphere, such as Earth , Mars, or the celestial sphere.
Degrees of freedom	In statistics, the number of degrees of freedom is the number of values in the final calculation of a statistic that are free to vary.
	Estimates of statistical parameters can be based upon different amounts of information or data.

Reduced cost	In linear programming, reduced cost, is the amount by which an objective function coefficient would have to improve (so increase for maximization problem, decrease for minimization problem) before it would be possible for a corresponding variable to assume a positive value in the optimal solution. It is the cost for increasing a variable by a small amount, i.e., the first derivative from a certain point on the polyhedron that constrains the problem. When the point is a vertex in the polyhedron, the variable with the most extreme cost, negatively for minimisation and positively maximisation, is sometimes referred to as the steepest edge.
Freedom	Freedom (often referred to as the Freedom app) is a software program designed to keep a computer user away from the Internet for up to eight hours at a time. It is described as a way to 'free you from distractions, allowing you time to write, analyze, code, or create.' The program was written by Fred Stutzman, a Ph.D student at the University of North Carolina at Chapel Hill. Freedom is no longer donationware, but, rather, is available for purchase for $10US. A five-time trial is available for free at http://macfreedom.com/
Optimality criterion	In statistics, an optimality criterion provides a measure of the fit of the data to a given hypothesis. The selection process is determined by the solution that optimizes the criteria used to evaluate the alternative hypotheses. The term has been used to identify the different criteria that are used to evaluate a phylogenetic tree and include maximum likelihood, Bayesian, maximum parsimony, and minimum evolution.
Iteration	Iteration means the act of repeating a process usually with the aim of approaching a desired goal or target or result. Each repetition of the process is also called an 'iteration,' and the results of one iteration are used as the starting point for the next iteration. Mathematics Iteration in mathematics may refer to the process of iterating a function i.e. applying a function repeatedly, using the output from one iteration as the input to the next.
Block	In computing (specifically data transmission and data storage), a block is a sequence of bytes or bits, having a nominal length (a block size). Data thus structured are said to be blocked. The process of putting data into blocks is called blocking.
PATH	PATH is an environment variable on Unix-like operating systems, DOS, OS/2, and Microsoft Windows, specifying a set of directories where executable programs are located. In general, each executing process or user session has its own PATH setting. Unix and Unix-like

Chapter 3. THE SIMPLEX METHOD

Breakthrough	Breakthrough is an abstract strategy board game invented by Dan Troyka in 2000 and made available as a Zillions of Games file (ZRF). It won the 2001 8x8 Game Design Competition, even though the game was originally played on a 7x7 board, as it is trivially extendible to larger board sizes. Rules The board is initially set up as shown on the right.
Exit	On many computer operating systems, a computer process terminates its execution by making an exit system call. More generally, an exit in a multithreading environment means that a thread of execution has stopped running. The operating system reclaims resources (memory, files, etc).
Reduced Vertical Separation Minima	Reduced Vertical Separation Minima is an aviation term used to describe the reduction of the standard vertical separation required between aircraft flying at levels between FL290 (29,000 ft). and FL410 (41,000 ft). from 2,000 feet to 1,000 feet (or between 8,900 metres and 12,500 metres from 600 metres to 300 metres in China).
Knapsack problem	The knapsack problem is a problem in combinatorial optimization: Given a set of items, each with a weight and a value, determine the number of each item to include in a collection so that the total weight is less than or equal to a given limit and the total value is as large as possible. It derives its name from the problem faced by someone who is constrained by a fixed-size knapsack and must fill it with the most valuable items. The problem often arises in resource allocation where there are financial constraints and is studied in fields such as combinatorics, computer science, complexity theory, cryptography and applied mathematics.

1. In linear programming, _____, is the amount by which an objective function coefficient would have to improve (so increase for maximization problem, decrease for minimization problem) before it would be possible for a corresponding variable to assume a positive value in the optimal solution. It is the cost for increasing a variable by a small amount, i.e., the first derivative from a certain point on the polyhedron that constrains the problem. When the point is a vertex in the polyhedron, the variable with the most extreme cost, negatively for minimisation and positively maximisation, is sometimes referred to as the steepest edge.

   a. Relaxation
   b. Reduced cost
   c. Rosenbrock function
   d. Self-concordant function

2. _____ is an aviation term used to describe the reduction of the standard vertical separation required between aircraft flying at levels between FL290 (29,000 ft). and FL410 (41,000 ft). from 2,000 feet to 1,000 feet (or between 8,900 metres and 12,500 metres from 600 metres to 300 metres in China).

   a. Standard Terminal Automation Replacement System
   b. .NET Messenger Service
   c. Reduced Vertical Separation Minima
   d. Powdered activated carbon treatment

3. In the field of mathematical optimization, _____ is a relaxation method which approximates a difficult problem of constrained optimization by a simpler problem. A solution to the relaxed problem is an approximate solution to the original problem, and provides useful information.

   The method penalizes violations of inequality constraints using a Lagrangian multiplier, which imposes a cost on violations.

   a. .NET Messenger Service
   b. Lagrangian relaxation
   c. Markov algorithm
   d. Normal form

4. . In mathematics, there are several meanings of degree depending on the subject.

   A _____(in full, a degree of arc, arc degree, or arcdegree), usually denoted by ° (the _____symbol), is a measurement of a plane angle, representing $\frac{1}{360}$ of a turn. When that angle is with respect to a reference meridian, it indicates a location along a great circle of a sphere, such as Earth , Mars, or the celestial sphere.

   a. Delta set
   b. Degree
   c. Dold manifold

5. _____ is an abstract strategy board game invented by Dan Troyka in 2000 and made available as a Zillions of Games file (ZRF). It won the 2001 8x8 Game Design Competition, even though the game was originally played on a 7x7 board, as it is trivially extendible to larger board sizes.

Rules

The board is initially set up as shown on the right.

a. Breakthru
b. Camelot
c. Breakthrough
d. Crosstrack

ANSWER KEY
Chapter 3. THE SIMPLEX METHOD

1. b
2. c
3. b
4. b
5. c

You can take the complete Chapter Practice Test

for Chapter 3. THE SIMPLEX METHOD
on all key terms, persons, places, and concepts.

Online 99 Cents

http://www.epub7.6.20492.3.cram101.com/

Use www.Cram101.com for all your study needs

including Cram101's online interactive problem solving labs in

chemistry, statistics, mathematics, and more.

CHAPTER OUTLINE: KEY TERMS, PEOPLE, PLACES, CONCEPTS

	Assignment
	Comparison
	Assignment problem
	Exit
	Action
	Space
	Redundancy
	Cutting stock problem

CHAPTER HIGHLIGHTS & NOTES: KEY TERMS, PEOPLE, PLACES, CONCEPTS

| Assignment | In computer programming, an assignment statement sets or re-sets the value stored in the storage location(s) denoted by a variable name. In most imperative computer programming languages, assignment statements are one of the basic statements. Common notations for the assignment operator are and . |
| Comparison | In computer programming, comparison of two data items is effected by the comparison operators typically written as:> (greater than)< (less than)>= (greater than or equal to)<= (less than or equal to)= or == (exactly equal to)!=, <>, ~= or /= (not equal to)

These operators produce the logical value or , depending on the result of the comparison. For example, in the pseudo-code

the statements following are executed only if the value of the variable 'a' is greater than 1 (i.e. when the logical value of is).

Some programming languages make a syntactical distinction between the 'equals' of assignment (e.g. |

Chapter 4. STARTING SOLUTION AND CONVERGENCE

Assignment problem	The assignment problem is one of the fundamental combinatorial optimization problems in the branch of optimization or operations research in mathematics. It consists of finding a maximum weight matching in a weighted bipartite graph. In its most general form, the problem is as follows:There are a number of agents and a number of tasks.
Exit	On many computer operating systems, a computer process terminates its execution by making an exit system call. More generally, an exit in a multithreading environment means that a thread of execution has stopped running. The operating system reclaims resources (memory, files, etc).
Action	In the Unified Modeling Language, an action is a named element that is the fundamental unit of executable functionality. The execution of an action represents some transformation or processing in the modeled system. An action execution represents the run-time behavior of executing an action within a specific behavior execution.
Space	In mathematics, a space is a set with some added structure. Mathematical spaces often form a hierarchy, i.e., one space may inherit all the characteristics of a parent space. For instance, all inner product spaces are also normed vector spaces, because the inner product induces a norm on the inner product space such that: $\|x\| = \sqrt{\langle x, x \rangle}$. Modern mathematics treats 'space' quite differently compared to classical mathematics.
Redundancy	In engineering, redundancy is the duplication of critical components or functions of a system with the intention of increasing reliability of the system, usually in the case of a backup or fail-safe. In many safety-critical systems, such as fly-by-wire and hydraulic systems in aircraft, some parts of the control system may be triplicated, which is formally termed triple modular redundancy (TMR). An error in one component may then be out-voted by the other two.
Cutting stock problem	The cutting-stock problem is an optimization problem, or more specifically, an integer linear programming problem. It arises from many applications in industry. Imagine that you work in a paper mill and you have a number of rolls of paper of fixed width waiting to be cut, yet different customers want different numbers of rolls of various-sized widths. How are you going to cut the rolls so that you minimize the waste (amount of left-overs)?

Solving this problem to optimality can be economically significant: a difference of 1% for a modern paper machine can be worth more than one million USD per year. Formulation and solution approaches

The standard formulation for the cutting-stock problem (but not the only one) starts with a list of m orders, each requiring q_j, j = 1,..m pieces. We then construct a list of all possible combinations of cuts (often called 'patterns'), associating with each pattern a positive integer variable x_i representing how many times each pattern is to be used. The linear integer program is

then:minimize $\sum_{i=1}^{n} c_i x_i$ subject to $\sum_{i=1}^{n} a_{ij} x_i \geq q_j, \qquad \forall j = 1, \ldots, m$ and

$x_i \geq 0$, integer

where a_{ij} is the number of times order j appears in pattern and c_i is the cost (often the waste) of pattern . The precise nature of the quantity constraints can lead to subtly different mathematical characteristics. The above formulation's quantity constraints are minimum constraints (at least the given amount of each order must be produced, but possibly more). When c_i = 1 the objective minimises the number of utilised master items and, if the constraint for the quantity to be produced is replaced by equality, it is called the bin packing problem. The most general formulation has two-sided constraints (and in this case a minimum-waste solution may consume more than the

minimum number of master items): $q_j \leq \sum_{i=1}^{n} a_{ij} x_i \leq Q_j, \qquad \forall j = 1, \ldots, m$

This formulation applies not just to one-dimensional problems. Many variations are possible, including one where the objective is not to minimise the waste, but to maximise the total value of the produced items, allowing each order to have a different value.

In general, the number of possible patterns grows exponentially as a function of m, the number of orders. As the number of orders increases, it may therefore become impractical to enumerate the possible cutting patterns.

An alternative approach uses delayed column-generation. This method solves the cutting-stock problem by starting with just a few patterns. It generates additional patterns when they are needed. For the one-dimensional case, the new patterns are introduced by solving an auxiliary optimization problem called the knapsack problem, using dual variable information from the linear program. The knapsack problem has well-known methods to solve it, such as branch and bound and dynamic programming. The Delayed Column Generation method can be much more efficient than the original approach, particularly as the size of the problem grows.

1. In the Unified Modeling Language, an action is a named element that is the fundamental unit of executable functionality. The execution of an action represents some transformation or processing in the modeled system. An _____execution represents the run-time behavior of executing an action within a specific behavior execution.

   a. Activity
   b. Actor
   c. Action
   d. Enterprise Distributed Object Computing

2. In computer programming, an _____statement sets or re-sets the value stored in the storage location(s) denoted by a variable name. In most imperative computer programming languages, _____statements are one of the basic statements. Common notations for the assignment operator are and .

   a. Expression
   b. Identifier
   c. Assignment
   d. Powdered activated carbon treatment

3. In computer programming, comparison of two data items is effected by the comparison operators typically written as:> (greater than)< (less than)>= (greater than or equal to)<= (less than or equal to)= or == (exactly equal to)!=, <>, ~= or /= (not equal to)

   These operators produce the logical value or , depending on the result of the comparison. For example, in the pseudo-code

   the statements following are executed only if the value of the variable 'a' is greater than 1 (i.e. when the logical value of is).

   Some programming languages make a syntactical distinction between the 'equals' of assignment (e.g. assigns the value 1 to the variable 'a') and the 'equals' of _____().

   a. Comparison of programming languages
   b. Constant
   c. Comparison
   d. Keyword

4. . The _____ is one of the fundamental combinatorial optimization problems in the branch of optimization or operations research in mathematics. It consists of finding a maximum weight matching in a weighted bipartite graph.

   In its most general form, the problem is as follows:There are a number of agents and a number of tasks.

   a. Ellipsoid method
   b. Assignment problem
   c. Dead code

5. On many computer operating systems, a computer process terminates its execution by making an _____ system call. More generally, an exit in a multithreading environment means that a thread of execution has stopped running. The operating system reclaims resources (memory, files, etc).

    a. Initng
    b. Overlay
    c. Dead code
    d. Exit

1. c
2. c
3. c
4. b
5. d

You can take the complete Chapter Practice Test

for Chapter 4. **STARTING SOLUTION AND CONVERGENCE**
on all key terms, persons, places, and concepts.

Online 99 Cents

http://www.epub7.6.20492.4.cram101.com/

Use www.Cram101.com for all your study needs

including Cram101's online interactive problem solving labs in

chemistry, statistics, mathematics, and more.

CHAPTER OUTLINE: KEY TERMS, PEOPLE, PLACES, CONCEPTS

	Comparison
	Lagrangian relaxation
	Assignment
	Sparse
	ELEMENTARY
	Matrix
	Block
	Hessenberg matrix
	Selection
	Lower bound
	Breakthrough
	Assignment problem
	Optimality criterion
	Knapsack problem

Comparison	In computer programming, comparison of two data items is effected by the comparison operators typically written as:> (greater than)< (less than)>= (greater than or equal to)<= (less than or equal to)= or == (exactly equal to)!=, <>, ~= or /= (not equal to)
	These operators produce the logical value or , depending on the result of the comparison. For example, in the pseudo-code
	the statements following are executed only if the value of the variable 'a' is greater than 1 (i.e. when the logical value of is).
	Some programming languages make a syntactical distinction between the 'equals' of assignment (e.g. assigns the value 1 to the variable 'a') and the 'equals' of comparison ().
Lagrangian relaxation	In the field of mathematical optimization, Lagrangian relaxation is a relaxation method which approximates a difficult problem of constrained optimization by a simpler problem. A solution to the relaxed problem is an approximate solution to the original problem, and provides useful information.
	The method penalizes violations of inequality constraints using a Lagrangian multiplier, which imposes a cost on violations.
Assignment	In computer programming, an assignment statement sets or re-sets the value stored in the storage location(s) denoted by a variable name. In most imperative computer programming languages, assignment statements are one of the basic statements. Common notations for the assignment operator are and .
Sparse	In computer science, Sparse is a tool designed to find possible coding faults in the Linux kernel. This static analysis tool differed from other such tools in that it was initially designed to flag constructs that were only likely to be of interest to kernel developers, e.g. mixing pointers to user address space and pointers to kernel address space.
	Sparse contains built-in checks for known problematic and a set of annotations designed to convey semantic information about types, such as what address space pointers point to, or what locks a function acquires or releases.
ELEMENTARY	In computational complexity theory, the complexity class ELEMENTARY of elementary recursive functions is the union of the classes in the exponential hierarchy. $$\begin{aligned} \text{ELEMENTARY} &= \text{EXP} \cup 2\text{EXP} \cup 3\text{EXP} \cup \cdots \\ &= \text{DTIME}(2^n) \cup \text{DTIME}(2^{2^n}) \cup \text{DTIME}(2^{2^{2^n}}) \cup \cdots \end{aligned}$$

The name was coined by László Kalmár, in the context of recursive functions and undecidability; most problems in it are far from elementary. Some natural recursive problems lie outside ELEMENTARY, and are thus NONELEMENTARY. Most notably, there are primitive recursive problems which are not in ELEMENTARY. We knowLOWER-ELEMENTARY $\subsetneq$ EXPTIME $\subsetneq$ ELEMENTARY $\subsetneq$ PR

Whereas ELEMENTARY contains bounded applications of exponentiation (for example, $O(2^{2^n})$), PR allows more general hyper operators (for example, tetration) which are not contained in ELEMENTARY.Definition

The definitions of elementary recursive functions are the same as for primitive recursive functions, except that primitive recursion is replaced by bounded summation and bounded product.

Matrix	In hot metal typesetting, a matrix is a mold for casting the letters known as sorts used in letterpress printing. In letterpress typography the matrix of one letter is inserted into the bottom of a hand mould, the mould is locked and molten type metal is poured into a straight-sided vertical cavity above the matrix. When the metal has cooled and solidified the mould is unlocked and a newly-cast metal sort is removed, ready for composition with other sorts.
Block	In computing (specifically data transmission and data storage), a block is a sequence of bytes or bits, having a nominal length (a block size). Data thus structured are said to be blocked. The process of putting data into blocks is called blocking.
Hessenberg matrix	In linear algebra, a Hessenberg matrix is one that is 'almost' triangular. To be exact, an upper Hessenberg matrix has zero entries below the first subdiagonal, and a lower Hessenberg matrix has zero entries above the first superdiagonal. They are named for Karl Hessenberg.
Selection	In relational algebra, a selection (sometimes called a restriction to avoid confusion with SQL's use of SELECT) is a unary operation written as $\sigma_{a\theta b}(R)$ or $\sigma_{a\theta v}(R)$ where:• a and b are attribute names• θ is a binary operation in the set $\{<, \leq, =, \geq, >\}$. v is a value constant• R is a relation The selection $\sigma_{a\theta b}(R)$ selects all those tuples in R for which θ holds between the a and the b attribute. The selection

$\sigma_{a\theta v}(R)$ selects all those tuples in R for which θ holds between the a attribute and the value v.

For an example, consider the following tables where the first table gives the relation $Person$, the second table gives the result of $\sigma_{Age \geq 34}(Person)$ and the third table gives the result of $\sigma_{Age=Weight}(Person)$.

Lower bound	In mathematics, especially in order theory, an upper bound of a subset S of some partially ordered set (P, ≤) is an element of P which is greater than or equal to every element of S. The term lower bound is defined dually as an element of P which is lesser than or equal to every element of S. A set with an upper bound is said to be bounded from above by that bound, a set with a lower bound is said to be bounded from below by that bound.
	A subset S of a partially ordered set P may fail to have any bounds or may have many different upper and lower bounds. By transitivity, any element greater than or equal to an upper bound of S is again an upper bound of S, and any element lesser than or equal to any lower bound of S is again a lower bound of S. This leads to the consideration of least upper bounds: (or suprema) and greatest lower bounds (or infima).
Breakthrough	Breakthrough is an abstract strategy board game invented by Dan Troyka in 2000 and made available as a Zillions of Games file (ZRF). It won the 2001 8x8 Game Design Competition, even though the game was originally played on a 7x7 board, as it is trivially extendible to larger board sizes.
	Rules
	The board is initially set up as shown on the right.
Assignment problem	The assignment problem is one of the fundamental combinatorial optimization problems in the branch of optimization or operations research in mathematics. It consists of finding a maximum weight matching in a weighted bipartite graph.
	In its most general form, the problem is as follows:There are a number of agents and a number of tasks.
Optimality criterion	In statistics, an optimality criterion provides a measure of the fit of the data to a given hypothesis. The selection process is determined by the solution that optimizes the criteria used to evaluate the alternative hypotheses.

Knapsack problem	The knapsack problem is a problem in combinatorial optimization: Given a set of items, each with a weight and a value, determine the number of each item to include in a collection so that the total weight is less than or equal to a given limit and the total value is as large as possible. It derives its name from the problem faced by someone who is constrained by a fixed-size knapsack and must fill it with the most valuable items. The problem often arises in resource allocation where there are financial constraints and is studied in fields such as combinatorics, computer science, complexity theory, cryptography and applied mathematics.

1. In computer programming, comparison of two data items is effected by the comparison operators typically written as:> (greater than)< (less than)>= (greater than or equal to)<= (less than or equal to)= or == (exactly equal to)!=, <>, ~= or /= (not equal to)

    These operators produce the logical value or , depending on the result of the comparison. For example, in the pseudo-code

    the statements following are executed only if the value of the variable 'a' is greater than 1 (i.e. when the logical value of is).

    Some programming languages make a syntactical distinction between the 'equals' of assignment (e.g. assigns the value 1 to the variable 'a') and the 'equals' of _____().

    a. Comparison
    b. Constant
    c. Generator
    d. Keyword

2. . In relational algebra, a _____(sometimes called a restriction to avoid confusion with SQL's use of SELECT) is a unary operation written as $\sigma_{a\theta b}(R)$ or $\sigma_{a\theta v}(R)$ where:• a and b are attribute names• θ is a binary operation in the set $\{<, \leq, =, \geq, >\}$.• v is a value constant• R is a relation

    The _____ $\sigma_{a\theta b}(R)$ selects all those tuples in R for which θ holds between the a and the b attribute.

    The _____ $\sigma_{a\theta v}(R)$ selects all those tuples in R for which θ holds between the a attribute and the value v .

For an example, consider the following tables where the first table gives the relation $Person$, the second table gives the result of $\sigma_{Age \geq 34}(Person)$ and the third table gives the result of $\sigma_{Age=Weight}(Person)$.

a. .NET Messenger Service
b. Cards
c. CAS Calc P11
d. Selection

3. In statistics, an _____ provides a measure of the fit of the data to a given hypothesis. The selection process is determined by the solution that optimizes the criteria used to evaluate the alternative hypotheses. The term has been used to identify the different criteria that are used to evaluate a phylogenetic tree and include maximum likelihood, Bayesian, maximum parsimony, and minimum evolution.

a. Optimality criterion
b. Pivot table
c. PlanMaker
d. Quattro Pro

4. In the field of mathematical optimization, _____ is a relaxation method which approximates a difficult problem of constrained optimization by a simpler problem. A solution to the relaxed problem is an approximate solution to the original problem, and provides useful information.

The method penalizes violations of inequality constraints using a Lagrangian multiplier, which imposes a cost on violations.

a. .NET Messenger Service
b. Motorola Mobility v. Apple Inc.
c. Lagrangian relaxation
d. Novelty

5. . The _____ is a problem in combinatorial optimization: Given a set of items, each with a weight and a value, determine the number of each item to include in a collection so that the total weight is less than or equal to a given limit and the total value is as large as possible. It derives its name from the problem faced by someone who is constrained by a fixed-size knapsack and must fill it with the most valuable items.

The problem often arises in resource allocation where there are financial constraints and is studied in fields such as combinatorics, computer science, complexity theory, cryptography and applied mathematics.

a. Least cost planning methodology
b. Leverage Point Modeling
c. Knapsack problem

ANSWER KEY
Chapter 5. SPECIAL SIMPLEX IMPLEMENTATIONS AND OPTIMALITY CONDITIONS

1. a
2. d
3. a
4. c
5. c

You can take the complete Chapter Practice Test

for Chapter 5. SPECIAL SIMPLEX IMPLEMENTATIONS AND OPTIMALITY CONDITIONS
on all key terms, persons, places, and concepts.

Online 99 Cents

http://www.epub7.6.20492.5.cram101.com/

Use www.Cram101.com for all your study needs

including Cram101's online interactive problem solving labs in

chemistry, statistics, mathematics, and more.

Breakthrough

Weak duality

Dual problem

Strong duality

Shadow price

Assignment

Assignment problem

Reduced Vertical Separation Minima

Optimality criterion

Lagrangian relaxation

Hungarian algorithm

Simultaneous equations

Sensitivity analysis

Matrix

Integer programming

Criss-cross algorithm

Chapter 6. DUALITY AND SENSITIVITY ANALYSIS

Breakthrough	Breakthrough is an abstract strategy board game invented by Dan Troyka in 2000 and made available as a Zillions of Games file (ZRF). It won the 2001 8x8 Game Design Competition, even though the game was originally played on a 7x7 board, as it is trivially extendible to larger board sizes. Rules The board is initially set up as shown on the right.
Weak duality	In applied mathematics, weak duality is a concept in optimization which states that the solution to the primal (minimization) problem is always greater than or equal to the solution to an associated dual problem. This is opposed to strong duality which only holds in certain cases. Weak duality states that the duality gap is always greater than or equal to 0.
Dual problem	In constrained optimization, it is often possible to convert the primal problem (i.e. the original form of the optimization problem) to a dual form, which is termed a dual problem. Usually 'dual problem' refers to the 'Lagrangian dual problem' but other dual problems are used, for example, the Wolfe dual problem and the Fenchel dual problem. The Lagrangian dual problem is obtained by forming the Lagrangian, using nonnegative Lagrangian multipliers to add the constraints to the objective function, and then solving for some primal variable values that minimize the Lagrangian.
Strong duality	Strong duality is a concept in optimization such that the primal and dual solutions are equivalent. This is as opposed to weak duality such that the primal problem has optimal value greater than the dual problem. Strong duality holds if and only if the duality gap is equal to 0.
Shadow price	In constrained optimization in economics, the shadow price is the instantaneous change per unit of the constraint in the objective value of the optimal solution of an optimization problem obtained by relaxing the constraint. In other words, it is the marginal utility of relaxing the constraint, or, equivalently, the marginal cost of strengthening the constraint. In a business application, a shadow price is the maximum price that management is willing to pay for an extra unit of a given limited resource.
Assignment	In computer programming, an assignment statement sets or re-sets the value stored in the storage location(s) denoted by a variable name. In most imperative computer programming languages, assignment statements are one of the basic statements. Common notations for the assignment operator are and .

Chapter 6. DUALITY AND SENSITIVITY ANALYSIS

Assignment problem	The assignment problem is one of the fundamental combinatorial optimization problems in the branch of optimization or operations research in mathematics. It consists of finding a maximum weight matching in a weighted bipartite graph.
	In its most general form, the problem is as follows:There are a number of agents and a number of tasks.
Reduced Vertical Separation Minima	Reduced Vertical Separation Minima is an aviation term used to describe the reduction of the standard vertical separation required between aircraft flying at levels between FL290 (29,000 ft). and FL410 (41,000 ft). from 2,000 feet to 1,000 feet (or between 8,900 metres and 12,500 metres from 600 metres to 300 metres in China).
Optimality criterion	In statistics, an optimality criterion provides a measure of the fit of the data to a given hypothesis. The selection process is determined by the solution that optimizes the criteria used to evaluate the alternative hypotheses. The term has been used to identify the different criteria that are used to evaluate a phylogenetic tree and include maximum likelihood, Bayesian, maximum parsimony, and minimum evolution.
Lagrangian relaxation	In the field of mathematical optimization, Lagrangian relaxation is a relaxation method which approximates a difficult problem of constrained optimization by a simpler problem. A solution to the relaxed problem is an approximate solution to the original problem, and provides useful information.
	The method penalizes violations of inequality constraints using a Lagrangian multiplier, which imposes a cost on violations.
Hungarian algorithm	The Hungarian method is a combinatorial optimization algorithm which solves the assignment problem in polynomial time and which anticipated later primal-dual methods. It was developed and published by Harold Kuhn in 1955, who gave the name 'Hungarian method' because the algorithm was largely based on the earlier works of two Hungarian mathematicians: Dénes Konig and Jeno Egerváry.
	James Munkres reviewed the algorithm in 1957 and observed that it is (strongly) polynomial. Since then the algorithm has been known also as Kuhn-Munkres algorithm or Munkres assignment algorithm. The time complexity of the original algorithm was $O(n^4)$, however Edmonds and Karp, and independently Tomizawa noticed that it can be modified to achieve an $O(n^3)$ running time. Ford and Fulkerson extended the method to general transportation problems. In 2006, it was discovered that Carl Gustav Jacobi had solved the assignment problem in the 19th century, and published posthumously in 1890 in Latin. Layman's explanation

Say you have three workers: Jim, Steve and Alan. You need to have one of them clean the bathroom, another sweep the floors & the third wash the windows. What's the best (minimum-cost) way to assign the jobs? First we need a matrix of the costs of the workers doing the jobs.

Then the Hungarian algorithm, when applied to the above table would give us the minimum cost it can be done with: Jim cleans the bathroom, Steve sweeps the floors and Alan washes the windows.

Simultaneous equations	In mathematics, simultaneous equations are a set of equations containing multiple variables. This set is often referred to as a system of equations. A solution to a system of equations is a particular specification of the values of all variables that simultaneously satisfies all of the equations.
Sensitivity analysis	Sensitivity analysis is the study of how the uncertainty in the output of a model (numerical or otherwise) can be apportioned to different sources of uncertainty in the model input. A related practice is uncertainty analysis which focuses rather on quantifying uncertainty in model output. Ideally, uncertainty and sensitivity analysis should be run in tandem.
Matrix	In hot metal typesetting, a matrix is a mold for casting the letters known as sorts used in letterpress printing. In letterpress typography the matrix of one letter is inserted into the bottom of a hand mould, the mould is locked and molten type metal is poured into a straight-sided vertical cavity above the matrix. When the metal has cooled and solidified the mould is unlocked and a newly-cast metal sort is removed, ready for composition with other sorts.
Integer programming	An integer programming problem is a mathematical optimization or feasibility program in which some or all of the variables are restricted to be integers. In many settings the term refers to integer linear programming, which is also known as mixed integer programming when some but not all the variables are restricted to be integers. Integer programming is NP-hard.
Criss-cross algorithm	In mathematical optimization, the criss-cross algorithm denotes a family of algorithms for linear programming. Variants of the criss-cross algorithm also solve more general problems with linear inequality constraints and nonlinear objective functions; there are criss-cross algorithms for linear-fractional programming problems, quadratic-programming problems, and linear complementarity problems. Like the simplex algorithm of George B.

1. In applied mathematics, _____ is a concept in optimization which states that the solution to the primal (minimization) problem is always greater than or equal to the solution to an associated dual problem. This is opposed to strong duality which only holds in certain cases.

   _____ states that the duality gap is always greater than or equal to 0.

   a. Weak duality
   b. Wolfe duality
   c. Mathematical optimization
   d. 2-opt

2. _____ is an abstract strategy board game invented by Dan Troyka in 2000 and made available as a Zillions of Games file (ZRF). It won the 2001 8x8 Game Design Competition, even though the game was originally played on a 7x7 board, as it is trivially extendible to larger board sizes.

   Rules

   The board is initially set up as shown on the right.

   a. Breakthrough
   b. Camelot
   c. Capture Go
   d. Crosstrack

3. _____ is a concept in optimization such that the primal and dual solutions are equivalent. This is as opposed to weak duality such that the primal problem has optimal value greater than the dual problem. _____ holds if and only if the duality gap is equal to 0.

   a. Subderivative
   b. Strong duality
   c. Successive linear programming
   d. Sum-of-squares optimization

4. _____ is the study of how the uncertainty in the output of a model (numerical or otherwise) can be apportioned to different sources of uncertainty in the model input. A related practice is uncertainty analysis which focuses rather on quantifying uncertainty in model output. Ideally, uncertainty and _____ should be run in tandem.

   a. Sensitivity analysis
   b. Servo bandwidth
   c. Servomechanism
   d. Setpoint

5. . In hot metal typesetting, a _____ is a mold for casting the letters known as sorts used in letterpress printing.

In letterpress typography the _____ of one letter is inserted into the bottom of a hand mould, the mould is locked and molten type metal is poured into a straight-sided vertical cavity above the _____. When the metal has cooled and solidified the mould is unlocked and a newly-cast metal sort is removed, ready for composition with other sorts.

a. Monitor proofing
b. The Museum of Printing
c. Matrix
d. Paper density

ANSWER KEY
Chapter 6. DUALITY AND SENSITIVITY ANALYSIS

1. a
2. a
3. b
4. a
5. c

You can take the complete Chapter Practice Test

for Chapter 6. DUALITY AND SENSITIVITY ANALYSIS
on all key terms, persons, places, and concepts.

Online 99 Cents

http://www.epub7.6.20492.6.cram101.com/

Use www.Cram101.com for all your study needs

including Cram101's online interactive problem solving labs in

chemistry, statistics, mathematics, and more.

Chapter 7. THE DECOMPOSITION PRINCIPLE

CHAPTER OUTLINE: KEY TERMS, PEOPLE, PLACES, CONCEPTS

	Lower bound
	Block
	Matrix
	Lagrangian relaxation
	Optimization
	Optimization problem
	Dual problem
	Assignment
	Assignment problem

CHAPTER HIGHLIGHTS & NOTES: KEY TERMS, PEOPLE, PLACES, CONCEPTS

Lower bound	In mathematics, especially in order theory, an upper bound of a subset S of some partially ordered set (P, ≤) is an element of P which is greater than or equal to every element of S. The term lower bound is defined dually as an element of P which is lesser than or equal to every element of S. A set with an upper bound is said to be bounded from above by that bound, a set with a lower bound is said to be bounded from below by that bound.
	A subset S of a partially ordered set P may fail to have any bounds or may have many different upper and lower bounds. By transitivity, any element greater than or equal to an upper bound of S is again an upper bound of S, and any element lesser than or equal to any lower bound of S is again a lower bound of S. This leads to the consideration of least upper bounds: (or suprema) and greatest lower bounds (or infima).
Block	In computing (specifically data transmission and data storage), a block is a sequence of bytes or bits, having a nominal length (a block size). Data thus structured are said to be blocked.

Chapter 7. THE DECOMPOSITION PRINCIPLE

Matrix	In hot metal typesetting, a matrix is a mold for casting the letters known as sorts used in letterpress printing. In letterpress typography the matrix of one letter is inserted into the bottom of a hand mould, the mould is locked and molten type metal is poured into a straight-sided vertical cavity above the matrix. When the metal has cooled and solidified the mould is unlocked and a newly-cast metal sort is removed, ready for composition with other sorts.
Lagrangian relaxation	In the field of mathematical optimization, Lagrangian relaxation is a relaxation method which approximates a difficult problem of constrained optimization by a simpler problem. A solution to the relaxed problem is an approximate solution to the original problem, and provides useful information. The method penalizes violations of inequality constraints using a Lagrangian multiplier, which imposes a cost on violations.
Optimization	In mathematics, computer science and economics, optimization, refers to choosing the best element from some set of available alternatives. In the simplest case, this means solving problems in which one seeks to minimize or maximize a real function by systematically choosing the values of real or integer variables from within an allowed set. This formulation, using a scalar, real-valued objective function, is probably the simplest example; the generalization of optimization theory and techniques to other formulations comprises a large area of applied mathematics.
Optimization problem	In mathematics and computer science, an optimization problem is the problem of finding the best solution from all feasible solutions. Optimization problems can be divided into two categories depending on whether the variables are continuous or discrete. An optimization problem with discrete variables is known as a combinatorial optimization problem.
Dual problem	In constrained optimization, it is often possible to convert the primal problem (i.e. the original form of the optimization problem) to a dual form, which is termed a dual problem. Usually 'dual problem' refers to the 'Lagrangian dual problem' but other dual problems are used, for example, the Wolfe dual problem and the Fenchel dual problem. The Lagrangian dual problem is obtained by forming the Lagrangian, using nonnegative Lagrangian multipliers to add the constraints to the objective function, and then solving for some primal variable values that minimize the Lagrangian.
Assignment	In computer programming, an assignment statement sets or re-sets the value stored in the storage location(s) denoted by a variable name. In most imperative computer programming languages, assignment statements are one of the basic statements.

Assignment problem	The assignment problem is one of the fundamental combinatorial optimization problems in the branch of optimization or operations research in mathematics. It consists of finding a maximum weight matching in a weighted bipartite graph. In its most general form, the problem is as follows:There are a number of agents and a number of tasks.

CHAPTER QUIZ: KEY TERMS, PEOPLE, PLACES, CONCEPTS

1. In computing (specifically data transmission and data storage), a block is a sequence of bytes or bits, having a nominal length (a _____ size). Data thus structured are said to be blocked. The process of putting data into blocks is called blocking.

    a. Werner Buchholz
    b. Datagram
    c. Field specification
    d. Block

2. In mathematics, computer science and economics, _____, refers to choosing the best element from some set of available alternatives.

    In the simplest case, this means solving problems in which one seeks to minimize or maximize a real function by systematically choosing the values of real or integer variables from within an allowed set. This formulation, using a scalar, real-valued objective function, is probably the simplest example; the generalization of _____ theory and techniques to other formulations comprises a large area of applied mathematics.

    a. Energy minimization
    b. Ellipsoid method
    c. Open Shop Scheduling
    d. Optimization

3. . In hot metal typesetting, a _____ is a mold for casting the letters known as sorts used in letterpress printing.

    In letterpress typography the _____ of one letter is inserted into the bottom of a hand mould, the mould is locked and molten type metal is poured into a straight-sided vertical cavity above the _____. When the metal has cooled and solidified the mould is unlocked and a newly-cast metal sort is removed, ready for composition with other sorts.

    a. Monitor proofing

b. The Museum of Printing

c. Matrix

d. Paper density

4. In mathematics, especially in order theory, an upper bound of a subset S of some partially ordered set (P, ≤) is an element of P which is greater than or equal to every element of S. The term _____ is defined dually as an element of P which is lesser than or equal to every element of S. A set with an upper bound is said to be bounded from above by that bound, a set with a _____ is said to be bounded from below by that bound.

A subset S of a partially ordered set P may fail to have any bounds or may have many different upper and _____s. By transitivity, any element greater than or equal to an upper bound of S is again an upper bound of S, and any element lesser than or equal to any _____ of S is again a _____ of S. This leads to the consideration of least upper bounds: (or suprema) and greatest _____s (or infima).

a. Lexicographical order

b. Better-quasi-ordering

c. Boolean prime ideal theorem

d. Lower bound

5. In the field of mathematical optimization, _____ is a relaxation method which approximates a difficult problem of constrained optimization by a simpler problem. A solution to the relaxed problem is an approximate solution to the original problem, and provides useful information.

The method penalizes violations of inequality constraints using a Lagrangian multiplier, which imposes a cost on violations.

a. Lagrangian relaxation

b. Flow

c. Human Terrain System

d. Know-net consortium

ANSWER KEY
Chapter 7. THE DECOMPOSITION PRINCIPLE

1. d
2. d
3. c
4. d
5. a

You can take the complete Chapter Practice Test

for Chapter 7. THE DECOMPOSITION PRINCIPLE
on all key terms, persons, places, and concepts.

Online 99 Cents

http://www.epub7.6.20492.7.cram101.com/

Use www.Cram101.com for all your study needs

including Cram101's online interactive problem solving labs in

chemistry, statistics, mathematics, and more.

Order

Decision problem

Optimization

Lagrangian relaxation

Space

Interior point method

Assignment

PATH

Matrix

Closed system

Factorization

Barrier function

Average-case complexity

Hungarian algorithm

Bidirectional search

Order	Order (subtitled 'A Journal on the Theory of Ordered Sets and its Applications') is a quarterly peer-reviewed academic journal on order theory and its applications, published by Springer Science+Business Media. It was founded in 1984 by University of Calgary mathematics professor Ivan Rival; as of 2010, its editor in chief is Dwight Duffus, the Goodrich C. White Professor of Mathematics & Computer Science at Emory University and a former student of Rival's. According to the Journal Citation Reports, the 2009 impact factor of Order is 0.408, placing it in the fourth quartile of ranked mathematics journals.
Decision problem	In computability theory and computational complexity theory, a decision problem is a question in some formal system with a yes-or-no answer, depending on the values of some input parameters. For example, the problem 'given two numbers x and y, does x evenly divide y?' is a decision problem. The answer can be either 'yes' or 'no', and depends upon the values of x and y.
Optimization	In mathematics, computer science and economics, optimization, refers to choosing the best element from some set of available alternatives. In the simplest case, this means solving problems in which one seeks to minimize or maximize a real function by systematically choosing the values of real or integer variables from within an allowed set. This formulation, using a scalar, real-valued objective function, is probably the simplest example; the generalization of optimization theory and techniques to other formulations comprises a large area of applied mathematics.
Lagrangian relaxation	In the field of mathematical optimization, Lagrangian relaxation is a relaxation method which approximates a difficult problem of constrained optimization by a simpler problem. A solution to the relaxed problem is an approximate solution to the original problem, and provides useful information. The method penalizes violations of inequality constraints using a Lagrangian multiplier, which imposes a cost on violations.
Space	In mathematics, a space is a set with some added structure. Mathematical spaces often form a hierarchy, i.e., one space may inherit all the characteristics of a parent space. For instance, all inner product spaces are also normed vector spaces, because the inner product induces a norm on the inner product space such that: $\|x\| = \sqrt{\langle x, x \rangle}.$

Interior point method	Interior point methods (also referred to as barrier methods) are a certain class of algorithms to solve linear and nonlinear convex optimization problems. The interior point method was invented by John von Neumann. Von Neumann suggested a new method of linear programming, using the homogeneous linear system of Gordan (1873) which was later popularized by Karmarkar's algorithm, developed by Narendra Karmarkar in 1984 for linear programming.
Assignment	In computer programming, an assignment statement sets or re-sets the value stored in the storage location(s) denoted by a variable name. In most imperative computer programming languages, assignment statements are one of the basic statements. Common notations for the assignment operator are and .
PATH	PATH is an environment variable on Unix-like operating systems, DOS, OS/2, and Microsoft Windows, specifying a set of directories where executable programs are located. In general, each executing process or user session has its own PATH setting. Unix and Unix-like On POSIX and Unix-like operating systems, the variable is specified as a list of one or more directory names separated by colon () characters.
Matrix	In hot metal typesetting, a matrix is a mold for casting the letters known as sorts used in letterpress printing. In letterpress typography the matrix of one letter is inserted into the bottom of a hand mould, the mould is locked and molten type metal is poured into a straight-sided vertical cavity above the matrix. When the metal has cooled and solidified the mould is unlocked and a newly-cast metal sort is removed, ready for composition with other sorts.
Closed system	The term closed system has different meanings in different contexts. In thermodynamics, a closed system can exchange energy (as heat or work), but not matter, with its surroundings. In contrast, an isolated system cannot exchange any of heat, work, or matter with the surroundings, while an open system can exchange all of heat, work and matter.
Factorization	In mathematics, factorization is the decomposition of an object (for example, a number, a polynomial, or a matrix) into a product of other objects, or factors, which when multiplied together give the original. For example, the number 15 factors into primes as 3×5, and the polynomial $x^2 - 4$ factors as $(x - 2)(x + 2)$. In all cases, a product of simpler objects is obtained.

Barrier function	In constrained optimization, a field of mathematics, a barrier function is a continuous function whose value on a point increases to infinity as the point approaches the boundary of the feasible region (Nocedal and Wright 1999). It is used as a penalizing term for violations of constraints. The two most common types of barrier functions are inverse barrier functions and logarithmic barrier functions.
Average-case complexity	Average-case complexity is a subfield of computational complexity theory that studies the complexity of algorithms on random inputs.
	The study of average-case complexity has applications in the theory of cryptography.
	Leonid Levin presented the motivation for studying average-case complexity as follows::
	Related topics•Probabilistic analysis of algorithms
	Literature
	The literature of average case complexity includes the following work:•Franco, John (1986), 'On the probabilistic performance of algorithms for the satisfiability problem', Information Processing Letters 23 (2): 103-106, doi:10.1016/0020-0190(86)90051-7 .•Levin, Leonid (1986), 'Average case complete problems', SIAM Journal on Computing 15 (1): 285-286, doi:10.1137/0215020 .•Flajolet, Philippe; Vitter, J. S. (August 1987), Average-case analysis of algorithms and data structures, Tech.
Hungarian algorithm	The Hungarian method is a combinatorial optimization algorithm which solves the assignment problem in polynomial time and which anticipated later primal-dual methods. It was developed and published by Harold Kuhn in 1955, who gave the name 'Hungarian method' because the algorithm was largely based on the earlier works of two Hungarian mathematicians: Dénes Konig and Jeno Egerváry.
	James Munkres reviewed the algorithm in 1957 and observed that it is (strongly) polynomial. Since then the algorithm has been known also as Kuhn-Munkres algorithm or Munkres assignment algorithm. The time complexity of the original algorithm was $O(n^4)$, however Edmonds and Karp, and independently Tomizawa noticed that it can be modified to achieve an $O(n^3)$ running time. Ford and Fulkerson extended the method to general transportation problems. In 2006, it was discovered that Carl Gustav Jacobi had solved the assignment problem in the 19th century, and published posthumously in 1890 in Latin. Layman's explanation
	Say you have three workers: Jim, Steve and Alan. You need to have one of them clean the bathroom, another sweep the floors & the third wash the windows. What's the best (minimum-cost) way to assign the jobs? First we need a matrix of the costs of the workers doing the jobs.

Bidirectional search	Bidirectional search is a graph search algorithm that finds a shortest path from an initial vertex to a goal vertex in a directed graph. It runs two simultaneous searches: one forward from the initial state, and one backward from the goal, stopping when the two meet in the middle. The reason for this approach is that in many cases it is faster: for instance, in a simplified model of search problem complexity in which both searches expand a tree with branching factor b, and the distance from start to goal is d, each of the two searches has complexity $O(b^{d/2})$ (in Big O notation), and the sum of these two search times is much less than the $O(b^d)$ complexity that would result from a single search from the beginning to the goal.

1. _____ is a graph search algorithm that finds a shortest path from an initial vertex to a goal vertex in a directed graph. It runs two simultaneous searches: one forward from the initial state, and one backward from the goal, stopping when the two meet in the middle. The reason for this approach is that in many cases it is faster: for instance, in a simplified model of search problem complexity in which both searches expand a tree with branching factor b, and the distance from start to goal is d, each of the two searches has complexity $O(b^{d/2})$ (in Big O notation), and the sum of these two search times is much less than the $O(b^d)$ complexity that would result from a single search from the beginning to the goal.

   a. Bidirectional search
   b. DSW algorithm
   c. Hash function
   d. Linear hashing

2. . _____ is a subfield of computational complexity theory that studies the complexity of algorithms on random inputs.

   The study of _____ has applications in the theory of cryptography.

   Leonid Levin presented the motivation for studying _____ as follows::

   Related topics•Probabilistic analysis of algorithms

   Literature

   The literature of average case complexity includes the following work:•Franco, John (1986), 'On the probabilistic performance of algorithms for the satisfiability problem', Information Processing Letters 23 (2): 103-106, doi:10.1016/0020-0190(86)90051-7 .•Levin, Leonid (1986), 'Average case complete problems', SIAM Journal on Computing 15 (1): 285-286, doi:10.1137/0215020 .•Flajolet, Philippe; Vitter, J. S. (August 1987), Average-case analysis of algorithms and data structures, Tech.

a. ESPACE

b. Advice

c. Average-case complexity

d. Alphabet

3. _____(subtitled 'A Journal on the Theory of Ordered Sets and its Applications') is a quarterly peer-reviewed academic journal on order theory and its applications, published by Springer Science+Business Media. It was founded in 1984 by University of Calgary mathematics professor Ivan Rival; as of 2010, its editor in chief is Dwight Duffus, the Goodrich C. White Professor of Mathematics & Computer Science at Emory University and a former student of Rival's.

According to the Journal Citation Reports, the 2009 impact factor of Order is 0.408, placing it in the fourth quartile of ranked mathematics journals.

a. Order dimension

b. Order

c. Order topology

d. Order-embedding

4. The term _____ has different meanings in different contexts.

In thermodynamics, a _____ can exchange energy (as heat or work), but not matter, with its surroundings. In contrast, an isolated system cannot exchange any of heat, work, or matter with the surroundings, while an open system can exchange all of heat, work and matter.

a. Coil-globule transition

b. Closed system

c. Compressed fluid

d. Compressibility

5. In computer programming, an _____statement sets or re-sets the value stored in the storage location(s) denoted by a variable name. In most imperative computer programming languages, _____statements are one of the basic statements. Common notations for the assignment operator are and .

a. Expression

b. Identifier

c. Assignment

d. Ordered subset expectation maximization

Visit Cram101.com for full Practice Exams

ANSWER KEY
Chapter 8. COMPLEXITY OF THE SIMPLEX ALGORITHM AND POLYNOMIAL-TIME ALGORITHMS

1. a
2. c
3. b
4. b
5. c

You can take the complete Chapter Practice Test

for Chapter 8. COMPLEXITY OF THE SIMPLEX ALGORITHM AND POLYNOMIAL-TIME ALGORITHMS
on all key terms, persons, places, and concepts.

Online 99 Cents

http://www.epub7.6.20492.8.cram101.com/

Use www.Cram101.com for all your study needs

including Cram101's online interactive problem solving labs in

chemistry, statistics, mathematics, and more.

CHAPTER OUTLINE: KEY TERMS, PEOPLE, PLACES, CONCEPTS

Lagrangian relaxation

Assignment

Assignment problem

Flow

Balance

PATH

Arborescence

CHAIN

Degree

Leaf node

Matrix

Incidence matrix

Simplex algorithm

Marginal cost

Exit

Iteration

Lower bound

Data

Data structure

_____ Antecedent

_____ Flow network

_____ Action

_____ CPLEX

_____ Variable splitting

CHAPTER HIGHLIGHTS & NOTES: KEY TERMS, PEOPLE, PLACES, CONCEPTS

Lagrangian relaxation	In the field of mathematical optimization, Lagrangian relaxation is a relaxation method which approximates a difficult problem of constrained optimization by a simpler problem. A solution to the relaxed problem is an approximate solution to the original problem, and provides useful information.

The method penalizes violations of inequality constraints using a Lagrangian multiplier, which imposes a cost on violations. |
| Assignment | In computer programming, an assignment statement sets or re-sets the value stored in the storage location(s) denoted by a variable name. In most imperative computer programming languages, assignment statements are one of the basic statements. Common notations for the assignment operator are and . |
| Assignment problem | The assignment problem is one of the fundamental combinatorial optimization problems in the branch of optimization or operations research in mathematics. It consists of finding a maximum weight matching in a weighted bipartite graph.

In its most general form, the problem is as follows:There are a number of agents and a number of tasks. |
| Flow | Flow is middleware software, which allows data integration specialists to connect disparate systems, whether they are on-premise, hosted or in the cloud; transforming and restructuring data as required between environments. |

	Flow functionality can be utilised for data integration projects, EDI and data conversion activities. Flow has been created by Flow Software Ltd in NZ and is available through a variety of partner companies or directly from Flow Software in NZ and Australia.
Balance	Balance is a simple but powerful generic TCP proxy with round robin load balancing and failover mechanisms. Its behaviour can be controlled at runtime using a simple command line syntax. Balance successfully runs at least on Linux(386), Linux(Itanium), FreeBSD, BSD/OS, Solaris, Cygwin, Mac-OS X, HP-UX and many more.
PATH	PATH is an environment variable on Unix-like operating systems, DOS, OS/2, and Microsoft Windows, specifying a set of directories where executable programs are located. In general, each executing process or user session has its own PATH setting. Unix and Unix-like On POSIX and Unix-like operating systems, the variable is specified as a list of one or more directory names separated by colon () characters.
Arborescence	In graph theory, an arborescence is a directed graph in which, for a vertex u called the root and any other vertex v, there is exactly one directed path from u to v. In other words, an arborescence is a directed, rooted tree in which all edges point away from the root. Every arborescence is a directed acyclic graph (DAG), but not every DAG is an arborescence.
CHAIN	The CECED Convergence Working Group has defined a new platform, called CHAIN (Ceced Home Appliances Interoperating Network), which defines a protocol for interconnecting different home appliances in a single multibrand system. It allows for control and automation of all basic appliance-related services in a home: e.g., remote control of appliance operation, energy or load management, remote diagnostics and automatic maintenance support to appliances, downloading and updating of data, programs and services (possibly from the Internet).
Degree	In mathematics, there are several meanings of degree depending on the subject. A degree (in full, a degree of arc, arc degree, or arcdegree), usually denoted by ° (the degree symbol), is a measurement of a plane angle, representing $\frac{1}{360}$ of a turn. When that angle is with respect to a reference meridian, it indicates a location along a great circle of a sphere, such as Earth , Mars, or the celestial sphere.
Leaf node	In computer science, a leaf node is a node of a tree data structure that has zero child nodes. Often, leaf nodes are the nodes farthest from the root node.

Chapter 9. MINIMAL-COST NETWORK FLOWS

Matrix	In hot metal typesetting, a matrix is a mold for casting the letters known as sorts used in letterpress printing.

In letterpress typography the matrix of one letter is inserted into the bottom of a hand mould, the mould is locked and molten type metal is poured into a straight-sided vertical cavity above the matrix. When the metal has cooled and solidified the mould is unlocked and a newly-cast metal sort is removed, ready for composition with other sorts. |
| Incidence matrix | In mathematics, an incidence matrix is a matrix that shows the relationship between two classes of objects. If the first class is X and the second is Y, the matrix has one row for each element of X and one column for each element of Y. The entry in row x and column y is 1 if x and y are related (called incident in this context) and 0 if they are not. There are variations. |
| Simplex algorithm | In mathematical optimization, Dantzig's simplex algorithm is a popular algorithm for linear programming. The journal Computing in Science and Engineering listed it as one of the top 10 algorithms of the twentieth century.

The name of the algorithm is derived from the concept of a simplex and was suggested by T. S. Motzkin. |
| Marginal cost | In economics and finance, marginal cost is the change in total cost that arises when the quantity produced changes by one unit. That is, it is the cost of producing one more unit of a good. If the good being produced is infinitely divisible, so the size of a marginal cost will change with volume, as a non-linear and non-proportional cost function includes the following:•variable terms dependent to volume,•constant terms independent to volume and occurring with the respective lot size,•jump fix cost increase or decrease dependent to steps of volume increase.

In practice the above definition of marginal cost as the change in total cost as a result of an increase in output of one unit is inconsistent with the calculation of marginal cost as MC=dTC/dQ for virtually all non-linear functions. |
| Exit | On many computer operating systems, a computer process terminates its execution by making an exit system call. More generally, an exit in a multithreading environment means that a thread of execution has stopped running. The operating system reclaims resources (memory, files, etc). |
| Iteration | Iteration means the act of repeating a process usually with the aim of approaching a desired goal or target or result. Each repetition of the process is also called an 'iteration,' and the results of one iteration are used as the starting point for the next iteration.

Mathematics |

Iteration in mathematics may refer to the process of iterating a function i.e. applying a function repeatedly, using the output from one iteration as the input to the next.

Lower bound

In mathematics, especially in order theory, an upper bound of a subset S of some partially ordered set (P, ≤) is an element of P which is greater than or equal to every element of S. The term lower bound is defined dually as an element of P which is lesser than or equal to every element of S. A set with an upper bound is said to be bounded from above by that bound, a set with a lower bound is said to be bounded from below by that bound.

A subset S of a partially ordered set P may fail to have any bounds or may have many different upper and lower bounds. By transitivity, any element greater than or equal to an upper bound of S is again an upper bound of S, and any element lesser than or equal to any lower bound of S is again a lower bound of S. This leads to the consideration of least upper bounds: (or suprema) and greatest lower bounds (or infima).

Data

In computer science, data is information in a form suitable for use with a computer. Data is often distinguished from programs. A program is a sequence of instructions that detail a task for the computer to perform.

Data structure

In computer science, a data structure is a particular way of storing and organizing data in a computer so that it can be used efficiently.

Different kinds of data structures are suited to different kinds of applications, and some are highly specialized to specific tasks. For example, B-trees are particularly well-suited for implementation of databases, while compiler implementations usually use hash tables to look up identifiers.

Antecedent

An antecedent is the first half of a hypothetical proposition.

Examples:•If P, then Q.

This is a nonlogical formulation of a hypothetical proposition. In this case, the antecedent is P, and the consequent is Q.•If X is a man, then X is mortal.

'X is a man' is the antecedent for this proposition.

Flow network

In graph theory, a flow network is a directed graph where each edge has a capacity and each edge receives a flow. The amount of flow on an edge cannot exceed the capacity of the edge.

Chapter 9. MINIMAL-COST NETWORK FLOWS

Action	In the Unified Modeling Language, an action is a named element that is the fundamental unit of executable functionality. The execution of an action represents some transformation or processing in the modeled system. An action execution represents the run-time behavior of executing an action within a specific behavior execution.
CPLEX	IBM ILOG CPLEX Optimization Studio (often informally referred to simply as CPLEX) is an optimization software package. In 2004, the work on CPLEX earned the first INFORMS Impact Prize.
	The CPLEX Optimizer was named for the simplex method as implemented in the C programming language, although today it provides additional methods for mathematical programming and offers interfaces other than just C. It was originally developed by Robert E. Bixby and was offered commercially starting in 1988 by CPLEX Optimization Inc., which was acquired by ILOG in 1997; ILOG was subsequently acquired by IBM in January 2009. CPLEX continues to be actively developed under IBM.
	The IBM ILOG CPLEX Optimizer solves integer programming problems, very large linear programming problems using either primal or dual variants of the simplex method or the barrier interior point method, convex and non-convex quadratic programming problems, and convex quadratically constrained problems (solved via Second-order cone programming, or SOCP).
Variable splitting	In applied mathematics and computer science, variable splitting is a decomposition method that relaxes a set of constraints.
	When the variable x appears in two sets of constraints, it is possible to substitute the new variables $x1$ in the first constraints and $x2$ in the second, and then join the two variables with a new 'linking' constraint, which requires that $x1=x2$.
	This new linking constraint can be relaxed with a Lagrange multiplier; in many applications, a Lagrange multiplier can be interpreted as the price of equality between $x1$ and $x2$ in the new constraint.
	For many problems, when the equality of the split variables is relaxed, then the system is decomposed, and each subsystem can be solved independently, at substantial reduction of computing time and memory storage.

1. In mathematics, an _____ is a matrix that shows the relationship between two classes of objects. If the first class is X and the second is Y, the matrix has one row for each element of X and one column for each element of Y. The entry in row x and column y is 1 if x and y are related (called incident in this context) and 0 if they are not. There are variations.

   a. Incidence matrix
   b. Index case
   c. Indicator bacteria
   d. Indices of deprivation 2004

2. PATH is an environment variable on Unix-like operating systems, DOS, OS/2, and Microsoft Windows, specifying a set of directories where executable programs are located. In general, each executing process or user session has its own _____ setting. Unix and Unix-like

   On POSIX and Unix-like operating systems, the variable is specified as a list of one or more directory names separated by colon () characters.

   a. Relvar
   b. PATH
   c. Singleton variable
   d. Static variable

3. In the field of mathematical optimization, _____ is a relaxation method which approximates a difficult problem of constrained optimization by a simpler problem. A solution to the relaxed problem is an approximate solution to the original problem, and provides useful information.

   The method penalizes violations of inequality constraints using a Lagrangian multiplier, which imposes a cost on violations.

   a. .NET Messenger Service
   b. .NET Show
   c. Lagrangian relaxation
   d. Common Language Runtime

4. In computer science, a _____ is a node of a tree data structure that has zero child nodes. Often, _____s are the nodes farthest from the root node. In the graph theory tree, a _____ is a vertex of degree 1 other than the root (except when the tree has only one vertex; then the root, too, is a leaf).

   a. Leaf node
   b. Directed algebraic topology
   c. Dold manifold
   d. Doomsday conjecture

5. The CECED Convergence Working Group has defined a new platform, called _____(Ceced Home Appliances Interoperating Network), which defines a protocol for interconnecting different home appliances in a single multibrand system.

It allows for control and automation of all basic appliance-related services in a home: e.g., remote control of appliance operation, energy or load management, remote diagnostics and automatic maintenance support to appliances, downloading and updating of data, programs and services (possibly from the Internet).

a. Coalition Warrior Interoperability Demonstration
b. Compatibility mode
c. Conceptual interoperability
d. CHAIN

ANSWER KEY
Chapter 9. MINIMAL-COST NETWORK FLOWS

---

1. a
2. b
3. c
4. a
5. d

You can take the complete Chapter Practice Test

for Chapter 9. MINIMAL-COST NETWORK FLOWS
on all key terms, persons, places, and concepts.

Online 99 Cents

http://www.epub7.6.20492.9.cram101.com/

Use **www.Cram101.com** for all your study needs

including **Cram101's online interactive problem solving labs in**

chemistry, statistics, mathematics, and more.

	Assignment
	Assignment problem
	Matrix
	CHAIN
	Flow
	Lagrangian relaxation
	Exit
	Hungarian algorithm
	Integer programming
	Reduced Vertical Separation Minima
	Dual problem
	PATH
	Stepping
	Stepping stone

Chapter 10. THE TRANSPORTATION AND ASSIGNMENT PROBLEMS

Assignment	In computer programming, an assignment statement sets or re-sets the value stored in the storage location(s) denoted by a variable name. In most imperative computer programming languages, assignment statements are one of the basic statements. Common notations for the assignment operator are and .
Assignment problem	The assignment problem is one of the fundamental combinatorial optimization problems in the branch of optimization or operations research in mathematics. It consists of finding a maximum weight matching in a weighted bipartite graph. In its most general form, the problem is as follows:There are a number of agents and a number of tasks.
Matrix	In hot metal typesetting, a matrix is a mold for casting the letters known as sorts used in letterpress printing. In letterpress typography the matrix of one letter is inserted into the bottom of a hand mould, the mould is locked and molten type metal is poured into a straight-sided vertical cavity above the matrix. When the metal has cooled and solidified the mould is unlocked and a newly-cast metal sort is removed, ready for composition with other sorts.
CHAIN	The CECED Convergence Working Group has defined a new platform, called CHAIN (Ceced Home Appliances Interoperating Network), which defines a protocol for interconnecting different home appliances in a single multibrand system. It allows for control and automation of all basic appliance-related services in a home: e.g., remote control of appliance operation, energy or load management, remote diagnostics and automatic maintenance support to appliances, downloading and updating of data, programs and services (possibly from the Internet).
Flow	Flow is middleware software, which allows data integration specialists to connect disparate systems, whether they are on-premise, hosted or in the cloud; transforming and restructuring data as required between environments. Flow functionality can be utilised for data integration projects, EDI and data conversion activities. Flow has been created by Flow Software Ltd in NZ and is available through a variety of partner companies or directly from Flow Software in NZ and Australia.
Lagrangian relaxation	In the field of mathematical optimization, Lagrangian relaxation is a relaxation method which approximates a difficult problem of constrained optimization by a simpler problem. A solution to the relaxed problem is an approximate solution to the original problem, and provides useful information.

Exit	On many computer operating systems, a computer process terminates its execution by making an exit system call. More generally, an exit in a multithreading environment means that a thread of execution has stopped running. The operating system reclaims resources (memory, files, etc).
Hungarian algorithm	The Hungarian method is a combinatorial optimization algorithm which solves the assignment problem in polynomial time and which anticipated later primal-dual methods. It was developed and published by Harold Kuhn in 1955, who gave the name 'Hungarian method' because the algorithm was largely based on the earlier works of two Hungarian mathematicians: Dénes Konig and Jeno Egerváry.
	James Munkres reviewed the algorithm in 1957 and observed that it is (strongly) polynomial. Since then the algorithm has been known also as Kuhn-Munkres algorithm or Munkres assignment algorithm. The time complexity of the original algorithm was $O(n^4)$, however Edmonds and Karp, and independently Tomizawa noticed that it can be modified to achieve an $O(n^3)$ running time. Ford and Fulkerson extended the method to general transportation problems. In 2006, it was discovered that Carl Gustav Jacobi had solved the assignment problem in the 19th century, and published posthumously in 1890 in Latin. Layman's explanation
	Say you have three workers: Jim, Steve and Alan. You need to have one of them clean the bathroom, another sweep the floors & the third wash the windows. What's the best (minimum-cost) way to assign the jobs? First we need a matrix of the costs of the workers doing the jobs.
	Then the Hungarian algorithm, when applied to the above table would give us the minimum cost it can be done with: Jim cleans the bathroom, Steve sweeps the floors and Alan washes the windows.
Integer programming	An integer programming problem is a mathematical optimization or feasibility program in which some or all of the variables are restricted to be integers. In many settings the term refers to integer linear programming, which is also known as mixed integer programming when some but not all the variables are restricted to be integers.
	Integer programming is NP-hard.
Reduced Vertical Separation Minima	Reduced Vertical Separation Minima is an aviation term used to describe the reduction of the standard vertical separation required between aircraft flying at levels between FL290 (29,000 ft). and FL410 (41,000 ft). from 2,000 feet to 1,000 feet (or between 8,900 metres and 12,500 metres from 600 metres to 300 metres in China).
Dual problem	In constrained optimization, it is often possible to convert the primal problem (i.e.

	the original form of the optimization problem) to a dual form, which is termed a dual problem. Usually 'dual problem' refers to the 'Lagrangian dual problem' but other dual problems are used, for example, the Wolfe dual problem and the Fenchel dual problem. The Lagrangian dual problem is obtained by forming the Lagrangian, using nonnegative Lagrangian multipliers to add the constraints to the objective function, and then solving for some primal variable values that minimize the Lagrangian.
PATH	PATH is an environment variable on Unix-like operating systems, DOS, OS/2, and Microsoft Windows, specifying a set of directories where executable programs are located. In general, each executing process or user session has its own PATH setting. Unix and Unix-like
	On POSIX and Unix-like operating systems, the variable is specified as a list of one or more directory names separated by colon () characters.
Stepping	Stepping is a designation used by Intel and AMD (or any semiconductor manufacturer) to identify how much the design of a microprocessor has advanced from the original design. The stepping is identified by a combination of a letter and a number.
	Typically, the first version of a microprocessor comes out with stepping A0. As design improvements occur, later versions are identified by changes in the letter and number.
Stepping stone	A stepping stone (StSt) is a type of computer security measure which consists of placing several logical security systems, used as authentication servers, in a serial disposition to emulate a physical narrow channel, analogous to a physical path formed by stepping stones used to cross a river. Using this system, it is possible to apply a granular control over each system acting as a 'stone', establishing different risk levels as so many systems which have been placed in the series.
	For example, to grant a user with access to an OpenSSH server, for executing an application in a high-security environment, we could put a front-end system such as a Sun Solaris with a Citrix Metaframe in the 1st security layer.

1. PATH is an environment variable on Unix-like operating systems, DOS, OS/2, and Microsoft Windows, specifying a set of directories where executable programs are located. In general, each executing process or user session has its own _____ setting. Unix and Unix-like

    On POSIX and Unix-like operating systems, the variable is specified as a list of one or more directory names separated by colon () characters.

    a. Relvar
    b. Sigil
    c. PATH
    d. Static variable

2. The _____ is one of the fundamental combinatorial optimization problems in the branch of optimization or operations research in mathematics. It consists of finding a maximum weight matching in a weighted bipartite graph.

    In its most general form, the problem is as follows: There are a number of agents and a number of tasks.

    a. Ellipsoid method
    b. Integer points in convex polyhedra
    c. Assignment problem
    d. Assembly

3. In computer programming, an _____ statement sets or re-sets the value stored in the storage location(s) denoted by a variable name. In most imperative computer programming languages, _____ statements are one of the basic statements. Common notations for the assignment operator are and .

    a. Expression
    b. Assignment
    c. Application domain
    d. Assembly

4. An _____ problem is a mathematical optimization or feasibility program in which some or all of the variables are restricted to be integers. In many settings the term refers to integer linear programming, which is also known as mixed _____ when some but not all the variables are restricted to be integers.

    _____ is NP-hard.

    a. Application domain
    b. Linear bottleneck assignment problem
    c. Linear programming relaxation
    d. Integer programming

5. . In hot metal typesetting, a _____ is a mold for casting the letters known as sorts used in letterpress printing.

In letterpress typography the _____ of one letter is inserted into the bottom of a hand mould, the mould is locked and molten type metal is poured into a straight-sided vertical cavity above the _____. When the metal has cooled and solidified the mould is unlocked and a newly-cast metal sort is removed, ready for composition with other sorts.

a. Monitor proofing
b. Matrix
c. Pad printing
d. Paper density

ANSWER KEY
Chapter 10. THE TRANSPORTATION AND ASSIGNMENT PROBLEMS

1. c
2. c
3. b
4. d
5. b

You can take the complete Chapter Practice Test

for Chapter 10. THE TRANSPORTATION AND ASSIGNMENT PROBLEMS
on all key terms, persons, places, and concepts.

Online 99 Cents

http://www.epub7.6.20492.10.cram101.com/

Use www.Cram101.com for all your study needs

including Cram101's online interactive problem solving labs in

chemistry, statistics, mathematics, and more.

Hungarian algorithm

Flow

Lagrangian relaxation

Breakthrough

Block

PATH

Line search

Search problem

CPLEX

Convex analysis

Interior point method

CHAPTER HIGHLIGHTS & NOTES: KEY TERMS, PEOPLE, PLACES, CONCEPTS

Hungarian algorithm	The Hungarian method is a combinatorial optimization algorithm which solves the assignment problem in polynomial time and which anticipated later primal-dual methods. It was developed and published by Harold Kuhn in 1955, who gave the name 'Hungarian method' because the algorithm was largely based on the earlier works of two Hungarian mathematicians: Dénes Konig and Jeno Egerváry. James Munkres reviewed the algorithm in 1957 and observed that it is (strongly) polynomial. Since then the algorithm has been known also as Kuhn-Munkres algorithm or Munkres assignment algorithm. The time complexity of the original algorithm was $O(n^4)$, however Edmonds and Karp, and independently Tomizawa noticed that it can be modified to achieve an

$O(n^3)$ running time. Ford and Fulkerson extended the method to general transportation problems. In 2006, it was discovered that Carl Gustav Jacobi had solved the assignment problem in the 19th century, and published posthumously in 1890 in Latin. Layman's explanation

Say you have three workers: Jim, Steve and Alan. You need to have one of them clean the bathroom, another sweep the floors & the third wash the windows. What's the best (minimum-cost) way to assign the jobs? First we need a matrix of the costs of the workers doing the jobs.

Then the Hungarian algorithm, when applied to the above table would give us the minimum cost it can be done with: Jim cleans the bathroom, Steve sweeps the floors and Alan washes the windows.

Flow	Flow is middleware software, which allows data integration specialists to connect disparate systems, whether they are on-premise, hosted or in the cloud; transforming and restructuring data as required between environments. Flow functionality can be utilised for data integration projects, EDI and data conversion activities. Flow has been created by Flow Software Ltd in NZ and is available through a variety of partner companies or directly from Flow Software in NZ and Australia.

Lagrangian relaxation	In the field of mathematical optimization, Lagrangian relaxation is a relaxation method which approximates a difficult problem of constrained optimization by a simpler problem. A solution to the relaxed problem is an approximate solution to the original problem, and provides useful information. The method penalizes violations of inequality constraints using a Lagrangian multiplier, which imposes a cost on violations.

Breakthrough	Breakthrough is an abstract strategy board game invented by Dan Troyka in 2000 and made available as a Zillions of Games file (ZRF). It won the 2001 8x8 Game Design Competition, even though the game was originally played on a 7x7 board, as it is trivially extendible to larger board sizes. Rules The board is initially set up as shown on the right.

Block	In computing (specifically data transmission and data storage), a block is a sequence of bytes or bits, having a nominal length (a block size). Data thus structured are said to be blocked. The process of putting data into blocks is called blocking.

PATH	PATH is an environment variable on Unix-like operating systems, DOS, OS/2, and Microsoft Windows, specifying a set of directories where executable programs are located. In general, each executing process or user session has its own PATH setting. Unix and Unix-like
	On POSIX and Unix-like operating systems, the variable is specified as a list of one or more directory names separated by colon () characters.
Line search	In optimization, the line search strategy is one of two basic iterative approaches to finding a local minimum $\mathbf{x}^*$ of an objective function $f : \mathbb{R}^n \to \mathbb{R}$. The other approach is trust region.
	The line search approach first finds a descent direction along which the objective function f will be reduced and then computes a step size that decides how far $\mathbf{x}$ should move along that direction.
Search problem	In computational complexity theory and computability theory, a search problem is a type of computational problem represented by a binary relation. If R is a binary relation such that field(R) $\subseteq$ Γ^+ and T is a Turing machine, then T calculates R if:•If x is such that there is some y such that R(x, y) then T accepts x with output z such that R(x, z) (there may be multiple y, and T need only find one of them)•If x is such that there is no y such that R(x, y) then T rejects x
	Intuitively, the problem consists in finding a structure y in an object x. An algorithm is said to solve the problem if it behaves in the following way: if at least one corresponding structure exists, then one occurrence of this structure is outputted; otherwise, the algorithm stops with an appropriate output ('Item not found' or any message of the like).
CPLEX	IBM ILOG CPLEX Optimization Studio (often informally referred to simply as CPLEX) is an optimization software package. In 2004, the work on CPLEX earned the first INFORMS Impact Prize.
	The CPLEX Optimizer was named for the simplex method as implemented in the C programming language, although today it provides additional methods for mathematical programming and offers interfaces other than just C. It was originally developed by Robert E. Bixby and was offered commercially starting in 1988 by CPLEX Optimization Inc., which was acquired by ILOG in 1997; ILOG was subsequently acquired by IBM in January 2009. CPLEX continues to be actively developed under IBM.
	The IBM ILOG CPLEX Optimizer solves integer programming problems, very large linear programming problems using either primal or dual variants of the simplex method or the barrier interior point method, convex and non-convex quadratic programming problems, and convex quadratically constrained problems (solved via Second-order cone programming, or SOCP).

Chapter 11. THE OUT-OF-KILTER ALGORITHM

Convex analysis	Convex analysis is the branch of mathematics devoted to the study of properties of convex functions and convex sets, often with applications in convex minimization, a subdomain of optimization theory.

A convex set is a set $C \subseteq X$, for some vector space X, such that for any $x, y \in C$ and $\lambda \in [0, 1]$ then $\lambda x + (1 - \lambda)y \in C$. Convex functions

A convex function is any extended real-valued function $f : X \to \mathbb{R} \cup \{\pm\infty\}$ which satisfies Jensen's inequality, i.e. for any $x, y \in X$ and any $\lambda \in [0, 1]$ then
$$f(\lambda x + (1 - \lambda)y) \leq \lambda f(x) + (1 - \lambda)f(y).$$

Equivalently, a convex function is any (extended) real valued function such that its epigraph
$$\{(x, r) \in X \times \mathbb{R} : f(x) \leq r\}$$

is a convex set. Convex conjugate

The convex conjugate of an extended real-valued (not necessarily convex) function
$f : X \to \mathbb{R} \cup \{\pm\infty\}$ is $f^* : X^* \to \mathbb{R} \cup \{\pm\infty\}$ where X^* is the dual
$$f^*(x^*) = \sup_{x \in X}\{\langle x^*, x \rangle - f(x)\}$$
space of X, and :pp.75-79Biconjugate

The biconjugate of a function $f : X \to \mathbb{R} \cup \{\pm\infty\}$ is the conjugate of the conjugate, typically written as $f^{**} : X \to \mathbb{R} \cup \{\pm\infty\}$.

Interior point method	Interior point methods (also referred to as barrier methods) are a certain class of algorithms to solve linear and nonlinear convex optimization problems.

The interior point method was invented by John von Neumann. Von Neumann suggested a new method of linear programming, using the homogeneous linear system of Gordan (1873) which was later popularized by Karmarkar's algorithm, developed by Narendra Karmarkar in 1984 for linear programming.

1. In optimization, the _____ strategy is one of two basic iterative approaches to finding a local minimum $\mathbf{x}^*$ of an objective function $f : \mathbb{R}^n \to \mathbb{R}$. The other approach is trust region.

The _____ approach first finds a descent direction along which the objective function f will be reduced and then computes a step size that decides how far $\mathbf{x}$ should move along that direction.

   a. Line search
   b. Linear matrix inequality
   c. Linear programming decoding
   d. Linear search problem

2. Flow is middleware software, which allows data integration specialists to connect disparate systems, whether they are on-premise, hosted or in the cloud; transforming and restructuring data as required between environments. Flow functionality can be utilised for data integration projects, EDI and data conversion activities. Flow has been created by _____ Software Ltd in NZ and is available through a variety of partner companies or directly from _____ Software in NZ and Australia.

   a. Human Terrain System
   b. Flow
   c. Knowledge broker
   d. Knowledge market

3. In computing (specifically data transmission and data storage), a block is a sequence of bytes or bits, having a nominal length (a _____ size). Data thus structured are said to be blocked. The process of putting data into blocks is called blocking.

   a. Werner Buchholz
   b. Datagram
   c. Block
   d. Frame

4. . The Hungarian method is a combinatorial optimization algorithm which solves the assignment problem in polynomial time and which anticipated later primal-dual methods. It was developed and published by Harold Kuhn in 1955, who gave the name 'Hungarian method' because the algorithm was largely based on the earlier works of two Hungarian mathematicians: Dénes Konig and Jeno Egerváry.

   James Munkres reviewed the algorithm in 1957 and observed that it is (strongly) polynomial. Since then the algorithm has been known also as Kuhn-Munkres algorithm or Munkres assignment algorithm. The time complexity of the original algorithm was $O(n^4)$, however Edmonds and Karp, and independently Tomizawa noticed that it can be modified to achieve an $O(n^3)$ running time. Ford and Fulkerson extended the method to general transportation problems. In 2006, it was discovered that Carl Gustav Jacobi had solved the assignment problem in the 19th century, and published posthumously in 1890 in Latin. Layman's explanation

Say you have three workers: Jim, Steve and Alan. You need to have one of them clean the bathroom, another sweep the floors & the third wash the windows. What's the best (minimum-cost) way to assign the jobs? First we need a matrix of the costs of the workers doing the jobs.

Then the _____, when applied to the above table would give us the minimum cost it can be done with: Jim cleans the bathroom, Steve sweeps the floors and Alan washes the windows.

a. Knapsack problem
b. Hungarian algorithm
c. Linear programming relaxation
d. Matroid intersection

5. _____ is an abstract strategy board game invented by Dan Troyka in 2000 and made available as a Zillions of Games file (ZRF). It won the 2001 8x8 Game Design Competition, even though the game was originally played on a 7x7 board, as it is trivially extendible to larger board sizes.

Rules

The board is initially set up as shown on the right.

a. Breakthru
b. Camelot
c. Breakthrough
d. Crosstrack

ANSWER KEY
Chapter 11. THE OUT-OF-KILTER ALGORITHM

1. a
2. b
3. c
4. b
5. c

You can take the complete Chapter Practice Test

for Chapter 11. THE OUT-OF-KILTER ALGORITHM
on all key terms, persons, places, and concepts.

Online 99 Cents

http://www.epub7.6.20492.11.cram101.com/

Use www.Cram101.com for all your study needs

including Cram101's online interactive problem solving labs in

chemistry, statistics, mathematics, and more.

CHAPTER OUTLINE: KEY TERMS, PEOPLE, PLACES, CONCEPTS

Flow

Lagrangian relaxation

Assignment

Assignment problem

Dual problem

Block

NP-complete

PATH

Shortest path problem

Arborescence

Lower bound

Data

Data structure

Matrix

Flow	Flow is middleware software, which allows data integration specialists to connect disparate systems, whether they are on-premise, hosted or in the cloud; transforming and restructuring data as required between environments. Flow functionality can be utilised for data integration projects, EDI and data conversion activities. Flow has been created by Flow Software Ltd in NZ and is available through a variety of partner companies or directly from Flow Software in NZ and Australia.
Lagrangian relaxation	In the field of mathematical optimization, Lagrangian relaxation is a relaxation method which approximates a difficult problem of constrained optimization by a simpler problem. A solution to the relaxed problem is an approximate solution to the original problem, and provides useful information. The method penalizes violations of inequality constraints using a Lagrangian multiplier, which imposes a cost on violations.
Assignment	In computer programming, an assignment statement sets or re-sets the value stored in the storage location(s) denoted by a variable name. In most imperative computer programming languages, assignment statements are one of the basic statements. Common notations for the assignment operator are and .
Assignment problem	The assignment problem is one of the fundamental combinatorial optimization problems in the branch of optimization or operations research in mathematics. It consists of finding a maximum weight matching in a weighted bipartite graph. In its most general form, the problem is as follows:There are a number of agents and a number of tasks.
Dual problem	In constrained optimization, it is often possible to convert the primal problem (i.e. the original form of the optimization problem) to a dual form, which is termed a dual problem. Usually 'dual problem' refers to the 'Lagrangian dual problem' but other dual problems are used, for example, the Wolfe dual problem and the Fenchel dual problem. The Lagrangian dual problem is obtained by forming the Lagrangian, using nonnegative Lagrangian multipliers to add the constraints to the objective function, and then solving for some primal variable values that minimize the Lagrangian.
Block	In computing (specifically data transmission and data storage), a block is a sequence of bytes or bits, having a nominal length (a block size). Data thus structured are said to be blocked. The process of putting data into blocks is called blocking.
NP-complete	In computational complexity theory, the complexity class NP-complete is a class of decision problems.

	A decision problem L is NP-complete if it is in the set of NP problems so that any given solution to the decision problem can be verified in polynomial time, and also in the set of NP-hard problems so that any NP problem can be converted into L by a transformation of the inputs in polynomial time.
	Although any given solution to such a problem can be verified quickly, there is no known efficient way to locate a solution in the first place; indeed, the most notable characteristic of NP-complete problems is that no fast solution to them is known.
PATH	PATH is an environment variable on Unix-like operating systems, DOS, OS/2, and Microsoft Windows, specifying a set of directories where executable programs are located. In general, each executing process or user session has its own PATH setting. Unix and Unix-like
	On POSIX and Unix-like operating systems, the variable is specified as a list of one or more directory names separated by colon () characters.
Shortest path problem	In graph theory, the shortest path problem is the problem of finding a path between two vertices (or nodes) in a graph such that the sum of the weights of its constituent edges is minimized. An example is finding the quickest way to get from one location to another on a road map; in this case, the vertices represent locations and the edges represent segments of road and are weighted by the time needed to travel that segment.
	There are several variations according to whether the given graph is undirected, directed, or mixed.
Arborescence	In graph theory, an arborescence is a directed graph in which, for a vertex u called the root and any other vertex v, there is exactly one directed path from u to v. In other words, an arborescence is a directed, rooted tree in which all edges point away from the root. Every arborescence is a directed acyclic graph (DAG), but not every DAG is an arborescence.
Lower bound	In mathematics, especially in order theory, an upper bound of a subset S of some partially ordered set (P, ≤) is an element of P which is greater than or equal to every element of S. The term lower bound is defined dually as an element of P which is lesser than or equal to every element of S. A set with an upper bound is said to be bounded from above by that bound, a set with a lower bound is said to be bounded from below by that bound.
	A subset S of a partially ordered set P may fail to have any bounds or may have many different upper and lower bounds. By transitivity, any element greater than or equal to an upper bound of S is again an upper bound of S, and any element lesser than or equal to any lower bound of S is again a lower bound of S.

Data	In computer science, data is information in a form suitable for use with a computer. Data is often distinguished from programs. A program is a sequence of instructions that detail a task for the computer to perform.
Data structure	In computer science, a data structure is a particular way of storing and organizing data in a computer so that it can be used efficiently.

Different kinds of data structures are suited to different kinds of applications, and some are highly specialized to specific tasks. For example, B-trees are particularly well-suited for implementation of databases, while compiler implementations usually use hash tables to look up identifiers. |
| Matrix | In hot metal typesetting, a matrix is a mold for casting the letters known as sorts used in letterpress printing.

In letterpress typography the matrix of one letter is inserted into the bottom of a hand mould, the mould is locked and molten type metal is poured into a straight-sided vertical cavity above the matrix. When the metal has cooled and solidified the mould is unlocked and a newly-cast metal sort is removed, ready for composition with other sorts. |

CHAPTER QUIZ: KEY TERMS, PEOPLE, PLACES, CONCEPTS

1. In hot metal typesetting, a _____ is a mold for casting the letters known as sorts used in letterpress printing.

   In letterpress typography the _____ of one letter is inserted into the bottom of a hand mould, the mould is locked and molten type metal is poured into a straight-sided vertical cavity above the _____. When the metal has cooled and solidified the mould is unlocked and a newly-cast metal sort is removed, ready for composition with other sorts.

   a. Matrix
   b. The Museum of Printing
   c. Pad printing
   d. Paper density

2. . In computational complexity theory, the complexity class _____ is a class of decision problems. A decision problem L is _____ if it is in the set of NP problems so that any given solution to the decision problem can be verified in polynomial time, and also in the set of NP-hard problems so that any NP problem can be converted into L by a transformation of the inputs in polynomial time.

Although any given solution to such a problem can be verified quickly, there is no known efficient way to locate a solution in the first place; indeed, the most notable characteristic of _____ problems is that no fast solution to them is known.

a. P versus NP problem
b. Paper bag problem
c. NP-complete
d. Parallel metaheuristic

3. In the field of mathematical optimization, _____ is a relaxation method which approximates a difficult problem of constrained optimization by a simpler problem. A solution to the relaxed problem is an approximate solution to the original problem, and provides useful information.

The method penalizes violations of inequality constraints using a Lagrangian multiplier, which imposes a cost on violations.

a. .NET Messenger Service
b. Lagrangian relaxation
c. Motion chart
d. Nameplate

4. The _____ is one of the fundamental combinatorial optimization problems in the branch of optimization or operations research in mathematics. It consists of finding a maximum weight matching in a weighted bipartite graph.

In its most general form, the problem is as follows:There are a number of agents and a number of tasks.

a. Ellipsoid method
b. Integer points in convex polyhedra
c. Assignment problem
d. Nameplate

5. PATH is an environment variable on Unix-like operating systems, DOS, OS/2, and Microsoft Windows, specifying a set of directories where executable programs are located. In general, each executing process or user session has its own _____ setting. Unix and Unix-like

On POSIX and Unix-like operating systems, the variable is specified as a list of one or more directory names separated by colon () characters.

a. Relvar
b. Sigil
c. Singleton variable
d. PATH

---

1. a
2. c
3. b
4. c
5. d

You can take the complete Chapter Practice Test

for Chapter 12. MAXIMAL FLOW, SHORTEST PATH, MULTICOMMODITY FLOW, & NETWORK SYNTHESIS PROBLEMS
on all key terms, persons, places, and concepts.

Online 99 Cents

http://www.epub7.6.20492.12.cram101.com/

Use www.Cram101.com for all your study needs

including Cram101's online interactive problem solving labs in

chemistry, statistics, mathematics, and more.

Other Cram101 e-Books and Tests

Want More?
Cram101.com...

Cram101.com provides the outlines and highlights of your textbooks, just like this e-StudyGuide, but also gives you the PRACTICE TESTS, and other exclusive study tools for all of your textbooks.

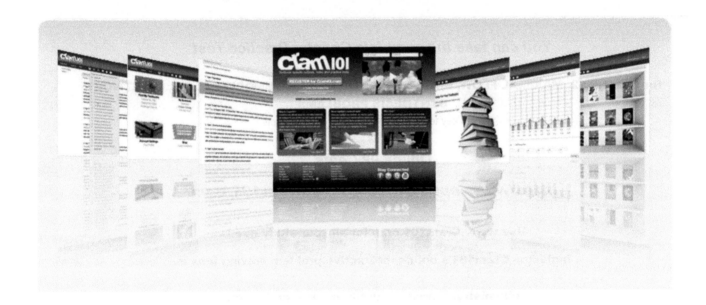

Learn More. *Just click*
http://www.cram101.com/

Other Cram101 e-Books and Tests

Lightning Source UK Ltd.
Milton Keynes UK
UKHW05f1146140318
319414UK00003B/104/P